Guns & Warriors

DTI Quips, Volume 2

John S Farnam

Guns & Warriors
DTI Quips, Volume 2
By John Farnam

Copyright © 2025 by John Farnam.
All rights reserved.

First Edition: September 2025
Published by Launch Pad Publications, LLC

Printed in the United States of America. No part of this book may be used or re-produced in any form or by any means or stored in a database or retrieval system without prior written permission of the publisher, except in the case of brief quotations embodied in critical articles and reviews. Making copies of any part of this book for any purpose other than your personal use violates United States copyright laws.

For more information or to book an event, visit
https://defense-training.com

ISBN - 979-8-9916724-1-2 (paperback)
ISBN - 979-8-9916724-3-6 (ebook)

Warning and Disclaimer
This book is sold as is, without warranty of any kind, either expressed or implied. While reasonable precaution has been taken in preparing this book, the author and Launch Pad Publications LLC. assume no responsibility for errors or omissions. Neither the author nor Launch Pad Publications LLC. shall have any liability to any person or entity concerning any loss or damage caused or alleged to be caused directly or indirectly by the instructions in this book. Further, neither the author nor Launch Pad Publications LLC assumes any responsibility for using or misusing the information or instructions contained herein.

To my wonderful sons and daughters, to my grandchildren, and most of all, to Vicki, who have all come together to make my life joyful and audacious!

About the Author

John S Farnam, president of Defense Training International LLC, has been teaching defensive firearms courses for almost 50 years. He is a combat veteran of the Vietnam War, a major (retired) in the U.S. Army Reserve and a police officer with many years of practical experience. Mr. Farnam is an acclaimed and prominent author, lecturer, expert witness and consultant. Through Defense Training International LLC, he teaches the latest defensive firearms techniques to police departments, federal and state agencies as well as foreign governments.

In 1996, John Farnam was awarded the prestigious "Tactical Advocate of the Year" title by the American Tactical Shooting Association in recognition of his commitment to presenting dependable tactical information to those attending his training programs. His years of experience in the military as well as law enforcement have allowed him to gain firsthand knowledge of what works most reliably in lethal force confrontations.

In April of 2009, he was inducted into Black Belt Magazine's "Living Legends." In November of 2011, he was elevated to the rank of Kyoshi Sensei within the American Marital Arts Association. In November of 2019, John was elevated to the rank of Sensei, Shodan 1st Degree Black Belt in Hojutsu-Ryu by its founder, Jeffrey Hall.

Contents

Part 1: Foundations & General Advice 1
   1. General Advice 3
   2. Travel Advice 23
   3. Women 27
   4. Gun Handling 31
   5. Techniques 39
   6. Tactics 43

Part 2: Case Studies & Incidents 55
   7. Accidental Discharges (ADs) 57
   8. Non-Police Gun Owner Incidents 59
   9. Police Incidents 63
  10. Officer Involved Shootings 73

Part 3: Weapons & Equipment 83
  11. Ammo 85
  12. Pistol Advice 89
  13. Pistol Reliability 97
  14. Pistol Problems 103
  15. Rifle Advice 107
  16. Rifle Reliability 119
  17. Shotgun Advice 123
  18. After Market Parts & Other Gear 129
  19. Hunting 137
  20. Duty Gear 143
  21. Training Tips 151

Part 4: Regional Experience 157
  22. Military 159
  23. Philippines 175
  24. South Africa 185
  25. Terrorism 201

Part 5: History 205
  26. America 207
  27. The World Wars 225
  28. Southeast Asia 229
  29. Other 233

Glossary 267
Index 273

Introduction

I've always viewed my role as a *"Defensive Firearms Instructor"* as involving not only teaching a set of psychomotor skills inherent to the correct and effective operation of defensive weapons but also teaching a companion philosophical overlay to complement those skills.

It's been said:

"A superior Gunman is best described as one who uses his superior judgement in an effort to avoid situations that would require a display of his superior skills"

During my lifetime, I for one, want to experience every good thing this life has to offer. I want to fall in love, have a family, travel, advance, accomplish, achieve, and inspire. However, sincerely pursuing none of the foregoing honorable goals is *"risk-free!"*

It has also been said:

"A ship's captain who really cares about the safety of his ship and crew would never leave port.

But that's not what ships are for!"

When instructing, I don't use the term, *"safe,"* when describing a procedure, nor a set of instructions, because the word *"safe"* implies a guarantee. When we use the term, *"safe,"* we imply that nothing could possibly go wrong, that all risk has been done away with.

What our students should be learning is that:

"Nothing good may be had without effort, without pain, nor without risk"

We teach students to effectively use, and live with, *"deadly weapons."* I don't think it is possible to handle deadly weapons "safely" I do think we can handle them carefully! Yet even *"careful/correct gun-handling"* does not guarantee that nothing bad will ever happen!

"Good tactics" does not mean *"taking no risk."* *"Good tactics"* means *"taking*

the best risk," with everything that term, *"risk,"* implies. Bad outcomes are still possible, no matter how careful, nor how *"safe,"* we try to be.

In the following pages you'll read about the exciting and diverse experiences of others, both contemporary and historical, and both good and bad.

"The essence of 'education' is learning from other peoples' mistakes."

Most mistakes you're ever likely to make during your tenure here on Earth have already been made. Learn about them, and then go forth boldly and audaciously,

... and make new mistakes!

John S Farnam, DTI
August 2025

Part 1: Foundations & General Advice

1. General Advice

21 Oct 2002

From a large gun retailer in the Midwest:

"Today, I had a couple come in the store looking for a concealable handgun that they could both use. The husband wanted something in 40S&W, saying his friends all told him the 9mm was 'largely ineffective.' I asked how many of his friends had ever been in a gunfight.

Of course, none of them had. I said that I was not trying to sell him on one caliber or another, but that, if both he and his wife were going to have to use the same pistol, both would have to be able to carry and use it comfortably.

I went on to say, 'There are no ineffective calibers, just ineffective shooters.' I indicated that both he and his wife would have to be trained to the point where their personal expertise was sufficient so as to make caliber irrelevant. After that, discussions about caliber would be largely extraneous.

I believe both he and his wife finally began to understand the undertaking they were contemplating. This is a discussion I have with all first-time gun buyers."

Lesson: ALL pistol cartridges are "largely ineffective!" "Effectiveness" comes from the shooter, not the gun.

24 Oct 2002

"Unintended consequences?" From an LEO friend in Baltimore.

"Drug dealers here have now realized that police in my city (Baltimore) have begun to enter cartridge cases found at the scene of our (daily) murders here into a state database for comparison. Of course, the effectiveness of this 'database' is highly dubious, but their universal response has nonetheless been to abandon autoloading pistols and instead carry and use revolvers exclusively, so they don't leave cartridge cases at the scene.

The PD, most of whose members don't even remember the 'revolver days,' has thus had to refamiliarize itself with S&W, Colt, and Ruger revolvers."

Lesson: No matter what laws are passed, criminals will quickly develop ways to work around them, in most cases, with scant inconvenience. Restrictive gun laws thus have no effect on crime but do have the effect of discouraging gun ownership among noncriminal citizens, which is, of course, their only real purpose.

17 Dec 2002

Latest from NJ:

"The state of New Jersey is passing a bill that would require all handguns sold after 2007 to be equipped with 'smart gun technology,' even though, at present, such technology is nonexistent.

In an effort to assist with the magical and precipitous appearance this currently nonexistent technology, the state has given millions in taxpayer dollars to the New Jersey Institute of Technology. NJIT's latest attempt (of many; all the rest have since been abandoned) is to develop a device that measures and recognizes the 'unique and distinct' position and pressure of a shooter's grip and trigger press. They assure us this device would recognize the characteristics of the programed shooter, thus allowing the gun to fire (at some point) after the recognition of the 'distinct signature' of bone structure and pressure produced when they grip is attained.

Questions like: Does this mean that I can never gain or lose weight? Does this mean that I can only hold the gun one way and in only one hand? Will it work when I'm wearing gloves? How many people can be programmed into one gun and how long will it take? Will all this add unacceptable bulk and weight to a gun that is designed to be carried concealed? What happens when the batteries go dead? Will it work in all foreseeable ambient temperatures, pressures, and humidity? Will the technology itself ever be reliable enough for 'emergency' equipment? Will this make defensive handguns prohibitively expensive?... have all remained casually and contemptuously unanswered. However, it is interesting to note that state law enforcement agencies, including the governor's bodyguard detail, will be exempt. We are told that the whole thing would be 'impractical' for them!

Those of us who are temporary stuck in the state of NJ are not impressed with this dubious and comically abstract 'technology', nor by the fact that our tax dollars are being spent on this fools' errand. As a result, we are all currently engaging in a desperate attempt by acquire as many 'dumb' handguns as we can possibly afford.

While all this is going on, our state's black bear population is rapidly getting out of hand. Attacks and other 'negative contacts' with black bears are becoming alarmingly commonplace. The obvious solution is to simply institute a hunting season. But, of course, the very notion of citizens actually benefitting is unthinkable for any good Democrat. So, our Game Department has received a grant to research a 'contraceptive' for female bears. No one has yet suggested how all these female bears are going to be persuaded to take 'the pill,' but, of course, no good bureaucrat ever worries about such things."

2025 Update: This stupid law never took effect! The technology still does not exist and probably never will.

29 Dec 2002

"Rabbit Chasing"

The term comes from the first scene in Alice in Wonderland. The heroine vainly chases an elusive rabbit, but never catches it.

At the poker table, the term, "rabbit chasing," refers to the situation where losing players ask the dealer to show them the card they might have gotten or ask the winning player, after all others have dropped out, what his hand was or if he was bluffing.

In poker etiquette, rabbit chasing is an unambiguous faux pas. Repeated instances of the practice will get the offending player invited to leave the table! At the poker table, good manners and personal honor are expected of all players, and ungraciously seeking information to which one is not entitled under the rules of the game is the sure sign of a bungling amateur or a cheater. It is also foolish! The information they think they want is actually damaging to them and to their game. It will do them far more harm than good.

The prophets of Old Testament all led miserable lives, and, in the end, most were homicide victims. Although the prospect (on the surface) seems attractive, only a fool would want to know the future.

Likewise, spending time and emotional energy wondering about what might have been will drive you to inaction, and maybe crazy too! Wondering what might have been is little more than looking for an excuse to lose.

In poker, as in fighting, winners spend their time and energy looking for a way to win. Losers spend their time looking for an excuse to lose. Both are usually successful. In poker, at the end of every hand, losers want to talk endlessly about what might have been. Winners just smile fearlessly and say, "Deal the cards!"

1 May 03

From a friend in AZ:

"Today's front-page lead, 'Gone Without a Trace'. Eight hundred registered sex offenders in Arizona have 'disappeared.' That is, they failed to check in as they're supposed to, so nobody knows where they are, nor what they're doing.

When questioned about this 'problem,' the supervisor of the Sex Offender Compliance Unit, Department of Public Safety said, 'We try to keep them in compliance, but you're always going to have a portion of the population that doesn't want to be tracked.' Well, that kind of feeble excuse sure

General Advice

builds my confidence in the 'Justice' System. How about you?

It gets worse. Incredibly, the State of Arizona is relatively good at this, sixth best in the nation in fact, with (only) six percent unaccounted for! California is missing 33%. Nevada 26%. Worst are Tennessee and Oklahoma at 50%! Combined, 77,000 of these convicted perverts are thumbing their collective noses at the 'System,' and, with virtually all state governments broke several times over, we all know that nothing is going to be done to improve the situation any time soon.

In years past, perverts were castrated or hanged. Today, they're slapped on the hand and then forgotten by politicians who are far more concerned with the next election than they ever will be about the safety of citizens.

Comment: But we're all assured that the 'System' is not broken and that none of us need to own a gun. The painful truth is that we are all on our own. No one is going to rescue us. No one 'cares' about us, and the cavalry isn't going to arrive in the nick of time. Individually, we have to be prepared to take care of business- any time, any place. Depending upon 'Justice System' for any species of 'protection' is (obviously) delusional.

11 May 03

"Hunches," "Feelings," "Lucky Charms," "Intuition," "ESP"?

It's all crap!

Both our lives and our craft are ruthlessly fact oriented. Indeed, so ruthless and unfair does life often seem, that many of us unwisely surrender to beliefs in ESP and the like, in an effort to cushion the impact of stark reality. When we do, we are immersed in an ocean of self-deception and ultimately do ourselves no good service! In simplest terms, sincere belief in ESP is a form of mental illness.

Casino games are literal manifestations of statical probability. However, gamblers love to talk about "lucky cards," "lucky dice," "winning streaks," "losing streaks," and the like. The casino business is profitable just because customers play "hunches," while the casino itself plays the odds! The fact is, low-probability events do occur. When they do, someone will always claim to have "seen it coming." Curiously, that claim always comes right after the event!

If you really believe "streaks" have any predictive value, look at the roulette wheel. Most casinos post the last ten results on an electronic bulletin board that is conspicuously situated above the wheel. Naive players are forever looking for "patterns" on the display. Unhappily, the only "pattern" you'll ever see is a classic bell curve! Indeed, casinos often keep records for months even years, just to make sure the wheel is "regular." They track millions, indeed billions, of spins. Their data shows exactly what any intelligent and rational person would expect: if red has come up eight times in a row, the ninth spin will come up red exactly as often as after any other sequence. At the craps table, I've seen the dice come up seven, five times in a row. The chances that the sixth throw will also be a seven is still one in six. There is no voodoo operating here, just statical probability in its purest form. When a seven is thrown when you have money on a number, you lose. You lose, no matter how "unfair" it seems, no matter what your "feelings" were, no matter how fervently you pray!

The laws of randomness, like to law of gravity, apply to everyone equally. We all need to stop "feeling" and start thinking. Your can "feel" any way you want. It will change nothing. "Feeling" is for losers! You can't predict the cards you're going to be delt.

Stop imagining that you can. Start concentrating on the one thing you can control: the way you play them. Victory dwells in the heart of the warrior, not in circumstances.

24 June 2003

Your "Lizard Brain"

In a recent conversation with good friend, Dave Grossman, Dave mentioned that he had recently talked with a gaggle of bearded, bespectacled psychiatrists (all with heavy, German accents). Dave was getting their advice on the differences between the human "front brain" and the "mid-brain." They had a number of terms for the "mid-brain," all with a minimum of six syllables and all difficult to pronounce. When Dave suggested to them the term, "mid-brain," they all nodded in wavering agreement that the term was probably adequately descriptive and that longer and more difficult terms would never see general use anyway.

What Dave, Gary Klugiewicz, and I all concur on is that lifesaving, psychomotor skills, intended to be used in an emergency must eventually filter from the frontal lobes (front brain), where they are first learned, into the mid-brain (primitive or "lizard" brain) if they are ever going to be accessible when one is in a hyper-stressful, crisis environment.

The frontal lobes is where our intellect dwells. Its precocious and elevated development separates us from lower forms of life. In one's frontal lobes lives discernment, understanding, and our ethical skeleton. However, the frontal lobes are also the residence of confusion, indecision, hesitation, and panic. The frontal lobes are never really quite sure of anything! The front brain is the "legislative branch" of our intelligence. The mid-brain is the "executive branch." The front brain works just fine when we are, at a leisurely pace, contemplating our navels, but, in a life-threatening emergency, a shrewd front brain wisely hands off operations to the mid-brain.

The mid-brain has no philosophy, no hesitation, and no regret. It knows only death, and life, and nothing in between! The mid-brain is never confused and never dithers. Its job is to get us out of this mess alive! It is poor at multitasking. It acts decisively and only does one thing at a time. It never apologizes, never looks back, and sheds no tears.

Unfortunately, the mid-brain is ignored in the training philosophy of many institutions. We do too much training "in the abstract." "In the abstract" is where all training must begin, because the front brain is the entry point for all information. Unhappily, that is where much of what passes for training also ends. As the student is gradually immersed in the training environment, stress levels must be increased so that important psychomotor skills begin to filter into the mid-brain. The mid-brain will only "know what to do" when the student has been "stress inoculated."

The hand-off from front brain to mid-brain must be seamless and immediate. The mid-brain has to "hit the ground running" if there is to be any chance that it can act in time to save your life. You need to "have a plan," and it must reside in the mid-brain. Unhelpful thoughts, swimming around in your front brain, must be jettisoned before they contaminate your mid-brain. This will mean endless repetitions under physical stress and anxiety.

Ultimately, your front brain will be of limited use during a crisis. In fact, it (and you, when you don't permit a hand-off to the mid-brain) will be little more than a blithering, dithering buffoon! When the hand-off to your mid-brain is smooth, authoritative, and timely, and your mid-brain has been well trained , it will know what to do and will act decisively to save your life.

Treat it well. Train it well!

3 July 2003

Saudi "Security Service" education, from a friend who is involved in contract training.

"Saudi 'Security Service' personnel all carry a S&W M10 revolver in 38Spl. On duty, it is carried unloaded. Speed loaders are forbidden (must watch out for that Western moral corruption). Madness, I know, but that is a direct order from the Saudi Ministry of Interior.

They also teach their personnel to fire warning shots (many of them), and, for the ultimate in lunacy, they shoot at the legs rather than center of visible mass. The foregoing is largely irrelevant anyway, because the men are all told that if they ever fire a shot from their revolver for any reason, they will go to prison. They are just 'beans,' and they know it. The Saudi government considers them utterly expendable.

Our 'training' course here in Virginia is seven days long. With religious breaks, it comes out to two days of actual training. Each student shoots a total of twenty-five actual rounds of ammunition. Upon graduation, none of them could fight their way out of a Riyadh whorehouse!"

Comment: This is yet another example of incompetence, elevated to an art form, complements of our Saudi "allies." Anyone who thinks the Saudis would or could put up a credible fight is dreaming.

10 July 2003

Letter to Justice O'Connor from an attorney:

"Dear Justice O'Connor:

In your University of Michigan Law School affirmative action decision, you have decreed that the government can pursue a fashionable, though ephemeral and nebulous, social policy by discriminating against me and my progeny on the basis of race.

The outcome of this case naturally was welcomed by the nation's elites. It is comforting and guilt assuaging, and its burdens are borne by others. Even the most doltish offspring of judges, academics, editors, government officials, business executives, etc., will always be admitted to the most prestigious institutions. The bright and hard-working offspring of poor Asians and lower-class Caucasians are the ones disadvantaged. Largely invisible and voiceless, they are ignored in decisions and the processes leading to them. Nothing has changed since the days that the same institutions had quotas limiting the number of Jews they would admit. The goal then also was to maintain the time's voguish concept of 'diversity.'

The reciprocal of your ruling that the provisions of the Fourteenth Amendment (and I cannot help but wonder what other Constitutional protections) are no longer available to me is the forfeiture by the government of any legitimate claim to loyalty, citizenship, nor other obligations from me. I do not consent to a government with the power to deny the equal protection of its laws to any of its citizens for any purpose, irrespective of how laudable and praiseworthy its goal may appear to be. Many conscientious citizens will, I believe, resist and subvert such a government when, as, and how they can."

Comment: The high court has decreed that racism is okay, so long as the trendy group benefits, and the "out of favor" group is punished. It is now just fine, according to the Supreme Court, to deny a person an opportunity solely because of the color of his skin. Now, watch each and every ethnic and religious group scramble to separate itself from the main body of Americans, as they all compete for "victim" status. It is a lawless decision. I, for one, am really worried about what these social activists will do with the Second Amendment and the rest of the Constitution!

13 July 2003

Response:

"Six years ago, a friend and his wife were having dinner with a local police officer and another couple. They didn't know the officer well but were happy to join him at his table when he invited them to do so. Dinner involved much wine drinking by all parties. Abruptly, the police officer pulled out a pistol (from concealment) and placed its muzzle to the head of the other male diner and held it there, as he continued to talk as if nothing had happened.

Everyone froze, except for the police officer, who continued talking. After a minute or two, he put the pistol away, and the four other diners all made expedient excuses and left the table, quickly exiting the restaurant as soon as they could. Everyone went home, and nothing ever came of the incident, legally or otherwise.

Sitting at dinner last night (six years later), after hearing the foregoing for the first time, I asked my friend (who has since become a lawyer and is currently the Public Defender for our county) if he wishes he would have done anything differently that night. I volunteered that a forcible disarm might have been in order, since I assume that, if someone pulls a gun and puts it to someone's head, that he intends to use it.

In true lawyerly fashion, my friend went on, ad nauseam, about how he could not possibly know this person's intent and that his actions, though threatening, were probably just his way of 'crying out for help.'

I begged the question and then asked him: 'If I had been the one threatened with a gun, and I had violently disarmed the perpetrator and subsequently shot him dead in order to end the threat and preserve my life and the lives of the other innocent people there, would you defend my actions in court?' His response was, once again, wavering. He said he didn't think defensive deadly force was ever really justified, that violent crime is not really a matter of personal choice on the part of the criminal, and that all crime, no matter how barbarous or cruel, is a direct result of "chemical imbalances."

With these people, there is always some cockeyed rationale for behavior that is unacceptable. The criminal always gets a pass. Decent people, who don't commit crimes, are always blamed. They are so paralyzed by evil that their only response is to deny its existence."

Comment: We tell our students to make decisions based on suspect capabilities, not suspect intent. People who try to guess at a threatening person's "intent" are already most of the way toward victimization.

A friend, who is an assistant district attorney, was recently in court prosecuting a defendant on a criminal charge. The judge asked if the Public Defender was ready to proceed, looking in the direction of the defendant's table.

My friend jumped up and said, "Excuse me, your honor, the attorney at the other table is not the 'Public Defender.' He is the 'Court-Appointed Representative of Indigent Defendants!'" There was a long pause. He continued, "I, sir, am the 'Public Defender!'"

Guns & Warriors: DTI Quips – Volume 2

Amen!

5 Aug 2003

At a Defensive Handgun Course in Colorado last week, Bob Brown of Soldier of Fortune Magazine decided to join us. I've known Bob for many years. In fact, we were in Vietnam at the same time.

Bob's magazine created quite a stir years ago when it was first published. Like Hugh Heffner, Bob has endured probing and disapproving looks from several administrations and, of course, the perpetually liberal media.

Bob is seventy now, but spry and every bit as enthusiastic as he ever was. He gets right out there and shoots with the rest of us. I asked him to lecture the class about some of his adventures. What a life he as lead! He has been to places I can't even pronounce.

Above all, Bob is a man of honor, an incorrigible purveyor of the truth, no matter how much others would like to ignore it. I am proud to call him my friend, and we all hope he goes on publishing forever.

Good show, Bob!

29 Aug 2003

On the Stealth Existence from a friend in GA:

"On the way back from an IDPA match today my car broke down, and I had to have it towed. The tow-truck driver was kind enough to drop me at my house en-route to dropping the car at my local mechanic's garage. Of course, I had to unload all my 'equipment' at the end of my driveway and carry it to the house.

Fortunately, my 'gun bag' is a toolbox from Home Depot. My 'ammo can' is a one-gallon paint can, and I carry my shotgun in a lacrosse racket case, as I learned from you. I was happy that I didn't have to unload a bunch of obvious gun stuff, such as hard rifle case, OD ammo can, and typical range bag at the end of my driveway in front of my neighbors, their kids, and the tow-truck driver. Instead, I looked like a guy on his way back from helping a friend with a home improvement project, which is, in fact, my usual cover story."

The Stealth Existence makes life tolerable and entails few sacrifices, except maybe ego-gratifying style concerns.

Lesson: The Stealth Existence must be regularly practiced by all of us. We don't want questions, nor curious looks. What we want to do is quietly "slip under the radar."

1 Sept 2003

A "stealth failure" from a friend in UT:

"A customer stopped at a 7-Eleven after leaving my shop and a subsequent desert shooting exercise with his prized DSA FAL, which he placed in his Chevy Blazer between the seats, sticking straight up in plain view. Someone at the 7-Eleven saw it and called the police. As he left the 7-Eleven, he discovered a police roadblock waiting for him. No one was hurt, but it was exciting. They cut him loose after several hours."

Lesson: The public is being incessantly conditioned by pinko politicians to believe all gun owners are criminals, that self-defense is immoral, and that being deliberately defenseless is a virtue. We are supposed to want to be victims, and when we refuse to go along with that mendacious line of thinking, we are vilified. All of us who are unapologetically not defenseless are thus increasingly viewed with suspicion and apprehension. Accordingly, our guns need to be out of sight and absent from casual conversation, always. The Stealth Existence offers the only real protection.

24 Sept 2003

From a friend in WI:

"Wisconsin's CCW bill will likely go off and quietly die this year. The repeated problem here (as in other states) has been the dismal political ineptitude of the legislation's proponents. Were it not for that, we would probably have a law in effect now.

Example: Last week, we had a high-profile, self-defense shooting here in Madison. It involved a local woman who deftly shot and killed two home invaders. Immediately seizing the opportunity, a showy local proponent of licensed concealed carry got on the evening news and, describing the woman as a saint, applauded her and declared that this is exactly why a CCW law is needed. The commentator (of course, no friend of gun owners) reminded him that the woman didn't require any kind of permit to defend herself in her own home. In addition, he pointed out that this "saintly" woman is actually a local crack dealer, and the inept decedents were two of her (apparently) unhappy customers. He made the interviewee look like a complete idiot, which he, unfortunately, is. Bungled interviews such as this surely put a bad face on our CCW campaign. The media, in fact, seeks out dithering buffoons like this whenever they want to discredit an issue they don't like.

Our local DA has decided not to charge the woman, despite the boisterous ranting of the decedents' relatives, who can't believe this woman had the audacity to actually defend herself with gunfire against the abortive attack of these two (thankfully departed) thugs. One presumes these same relatives would be standing solidly behind their criminal kin at the murder trials, had the home invasion not been successfully repelled."

Comment: As Second Amendment advocates, we must always be well spoken and have our facts straight. Some of us make poor spokesmen, no matter how sincere we are. Those of us in that category need to leave television interviews to others.

As Lincoln once put it, *"Better to remain silent and be thought a fool, than open your mouth and remove all doubt!"*

1 Oct 2003

The 6th Commandment

General Advice

Many friends and students have been troubled by the admonition in the Bible's Old Testament Sixth Commandment, ie: "Thou shall not kill." It is time this confusion be put to rest. Research through Strong's Exhaustive Concordance and several different translations of The Old Testament is helpful.

The clause in question is found in the book of Exodus, chapter twenty, verse thirteen. In the King James' Version (KJV), that verse is translated, "Thou shall not kill." In the days when this was written, animals were, of course, killed for food, and there is nowhere in scripture any instructions that this practice be stopped. So, we must examine the Hebrew word that is translated "kill" in the KJV in order to clarify exactly what is meant. Incidentally, it is only in the KJV and several other (old) translations that the word is translated "kill". In the Moffatt Translation, the verse is rendered, "You shall not murder". In The Living Bible, it is rendered, "You must not murder".

The Hebrew word translated to the English word "kill" in the KJV is "ratsach" (pronounced raw-tsakh'). It literally means, "to dash to pieces", and it is never used in context with animals. It refers exclusively to people. In fact, this particular word is used only several other times in The Old Testament, and each time it refers to homicide. A different word is used to describe the killing of animals for food or sacrifice. It is "shachat" (pronounced shaw-khat').

For example:

Numbers 35:15-16:

"Anyone who kills a person by accident may take sanctuary in them (your towns). But, if he (deliberately) struck the person with an iron tool, so that he died; the man is a murderer; the murderer must be put to death without fail."

The word here (ratsach) translated "murderer" is the same one translated "kill" back in Exodus. Apparently, KJV translators were not particularly consistent, unless it fit their agenda.

Continuing, Numbers 35:20-21:

"Also, if he pushes a person, because he hates him, or hides and throws something at him, so that he dies, or maliciously strikes him until he dies, the man who struck the blow must be put to death without fail; he is a murderer."

Again, the same word, "ratsach" is translated "murderer".

The word used to describe a public execution of a criminal, such as in the verses quoted above as "to put to death" is "nathan muwth" (naw-than' mooth). "Nathan" is a form of the verb, to do. It usually means, "to put," or "to make." "Muwth" is a form of the verb, "to die," and it refers to the state of death.

The word translated "kills" in the verse above, where the killing was an accident is, "nakah" (naw-kaw'). It means to commit a homicide, as does ratsach, but it is a softer word and is usually used to describe an accidental homicide, rather than a murder. There is no separate Hebrew word used to distinguish a deliberate, but justifiable homicide from a deliberate, but unjustifiable one. Ratsach is customarily used in both instances, much like our word, "homicide".

However, certain homicides are clearly identified as justifiable:

Numbers 35:26-27:

"If the murderer ever goes outside the bounds of the town of refuge where he has taken sanctuary and is caught by the avenger outside the bounds, then the avenger may kill (ratsach) the murderer without incurring any guilt."

From the foregoing, I think it is safe to say that Exodus 20:13 is correctly translated, "You shall not murder". "You shall not kill", is an obvious mistranslation, because it is not specific and engenders confusion, particularly among grasseaters who use it to rationalize their own willful unpreparedness and cowardice.

8 Oct 2003

An LDS friend reminded me of this quotation from Joseph Smith in the Mormon Church's Doctrine and Covenants, Section 134, Verse 11:

"We believe that men should appeal to the civil law for redress of all wrongs and grievances, where personal abuse is inflicted or the right of property or character infringed, where such laws exist as will protect the same; but we believe that all men are justified in defending themselves, their friends, and property, and the government, from the unlawful assaults and encroachments of all persons in times of exigency, where immediate appeal cannot be made to the laws, and relief afforded."

Curiously, the foregoing is exactly what we teach today.

Mormons got it right!

15 Oct 2003

Matching the Mission with the Gun:

I try my best to discourage it, but, at every Urban Rifle Course, at least one student will bring an M1A (sometimes even a M70 bolt gun) with a mammoth, high-magnification, close-eye-relief scope, usually 3-9X variable. In addition, the rifle itself has invariably been "accurized" to the point where it is tight, temperamental, and ammunition sensitive. Nothing inherently wrong with this equipment. It's just unsuitable to our mission!

This is what we all need to understand: We carry pistols as a way to deal with unexpected threats. As such, carry pistols need to be slim, slick, smooth, short, and easy and convenient to carry concealed. In addition, a carry pistol needs to have an adequate reserve of ammunition and be powerful enough to stop a fight even when several opponents must be neutralized quickly. Defensive pistols, indeed, all defensive firearms, also need to be loose enough to continue to operate reliably in spite of exposure to sweat, grit, lint and continuous neglect. No pistol is going to be nearly as effective (as accurate, as powerful) as a rifle or shotgun, but, unlike rifles and shotguns, we can have a pistol on our person (concealed) nearly all the time. Pistols are convenient, not particularly effective.

Utility/defensive rifles and shotguns cannot be carried on the person concealed (in most cases), but we keep them nearby as a means of dealing with expected threats, albeit threats that may still come at any time and from any direction. Like pistols, utility/defensive rifles and shotguns thus must be thought of as reactive/defensive weapons. Sometimes, rifles and shotguns must be employed at what are normally considered pistol ranges, where retention is a critical issue. Hence, defensive rifles and shotguns must be short, slick, and handy, but robust in much the same way a defensive pistol must be robust. In my opinion, for quick, reactive deployment, iron sights, red-dot optics or LPVOs (low-power, variable optics) are best for most people. Rifles and shotguns are effective, not particularly convenient, even in the best configurations.

Given the above, we must face the fact that any suitable, defensive rifle or pistol is never going to yield better than mediocre accuracy, at least according to the standards of competitive target shooters. Accuracy and reliability are mutually antagonistic. Of course, the inherent accuracy of any good, utility rifle is more than sufficient for the purpose to which we're putting it. Greater accuracy is certainly possible, but tighten a gun up to yield great precision, and you turn it into a moody, temperamental prima donna, not something you would want to protect your life. I am happy to give up superfluous accuracy in exchange for dependable reliability.

Finally, we use hyper-accurate, scoped rifles as a way of dealing with threats that are not only expected but are also known and identified, as in the case in a military context. These weapons are not "reactive" and thus are not particularly suitable for dealing with unknown and unlocated threats, be they expected or unexpected. These are temperamental, sniper rifles, requiring anal maintenance and gentle treatment. They do have their place in military sniping, just not in domestic, defensive/reactive shooting. People bring them to Urban Rifle courses, not because they're suitable, but because their owners want to show them off and look good.

General Advice

Tiny shot groups are always impressive, at any range, providing that "impressing people" is all you want to do.

If you're serious about defensive shooting, stay away from any rifle (or pistol) that is marked "Target" or "Match." Accept the fact that, to stay alive long enough to finish the fight, you're going to have to deploy your weapon quickly and shoot decisively, perhaps at multiple targets at various ranges, with sufficient accuracy to stop the fight, and sufficient speed to allow you to seize and hold the agenda.

We have to equip ourselves to fight, not play games!

21 Dec 2003

On shooting skills, from a friend and student:

"Today, Carol and I went to a local bowling alley to pick up our daughter from a birthday party. While we were waiting, Carol suggested we play one of the shooting games in the arcade.

We engaged in a duel match. Both of us had to fight off the thundering hordes of invading monsters. Funny, she beat me badly! Why? I haven't shot in a bit (a sore point with me) and got sloppy immediately. I didn't use the weapon's sights. I just pointed and yanked the trigger, assuming, hell, its just a game. After all, I'm a 'good shot.' I've been shooting for years. How can I fail?

I failed! Carol, on the other hand, lacked my arrogance and overconfidence. She only had the frame of reference from Vicki's instruction. Thus, while she was slower to engage, she used her sights, pressed the trigger carefully, tracked the target, and hit consistently.

Ultimately, the electronic score keeper revealed that I had engaged more targets, shot more bullets, but had far fewer hits than her. She fired slower and more deliberately but didn't miss at all. Appalled, I immediately challenged her to a rematch. She reminded me, 'Had this been a real fight, you wouldn't get a second chance!

It's 'back to the basics' for me."

Lesson: It can't be said too often: Accurate shooting wins fights. Sloppy shooting is for losers.

"Never do your enemy a 'minor injury'"

Machiavelli

2. Travel Advice

18 Oct 2002

Air travel advice from a friend who does even more than me:

1. Nobody questions canes. They get x-rayed, but that is about it.

2. Nobody questions small flashlights

3. I travel with books, including a small bible. None are ever examined, except the Bible, which is always examined.

18 Dec 2002

An individual was traveling by air last week and had a connection at Chicago's O'Hare Airport. He was stopped by the new federal security folks and informed that he had been selected for a random search of his checked baggage. Officers from the Chicago Police Department came along. His checked luggage was examined and found to contain a sheathed knife. I don't know how big it was or what brand. In any event, the feds were indifferent to the knife. He was told by the federal officer that it was okay to transport knifes in checked baggage. The examination was completed, and he was sent on his way.

"Not so fast," said the CPD officers. They arrested him, saying that, notwithstanding the federal rules on transport of knives in checked baggage, the City of Chicago considered this knife to be a "concealed weapon," in violation of city ordinances.

This person, of course, missed his flight. The case is still pending in Chicago.

The lesson here is that overlapping and contradictory laws that apply simultaneously make compliance or even knowing that you are or are not complying, impossible. In the government's ongoing war on all forms of privacy, authorities now have access to areas previously unexposed and considered private.

Several months ago, in a similar incident at O'Hare that involved me, I personally found the CPD guys to be pretty cool. At that time, they had no problem with knives and a number of other items I had in checked baggage. The above incident may have involved a young officer still in "academy mode." I don't know those details.

With the world situation the way it is, we must all reexamine our personal profile. The "stealth existence" is going to be the order of the day, and we all need to be experts at it!

31 Mar 2003

Air Travel Problems:

I just talked with a friend this morning who has a bad experience at Houston International Airport. He was flying United from Houston to Chicago on Friday of last week. He properly declared his pistol in his checked baggage. He also had two charged magazines (1911, 45ACP). The gun (unloaded) was in a locked box inside the suitcase. The magazines were in a side compartment of the same suitcase.

The United ticket agent was in full-Nazi mode, and the federal TSA agents were as terrified of her as was everyone else, including her boss. She growled several times that she "just hates guns." It was painfully obvious that she hates men too! When the gun was inspected, she personally removed all rounds from both magazines and announced (loud enough for everyone to hear) that they would be destroyed. No amount of quoting from the airline's own regulations would dissuade her.

My friend demanded to see a uniformed police officer. An HPD sergeant arrived, agreed with my friend that the ammunition was fine in magazines and that he was doing nothing wrong, but then added, "United owns this airport, and there is nothing I can do."

My friend is filing a complaint with both United and TSA.

Lesson: This ticket agent, reflecting the antigun attitude of United and

Travel Advice

most other airlines, brings her gun-hating attitude to work with her. She patently believes gun owners need to be harassed out of existence.

Being in full compliance with all regulations is no guarantee that you will not be harassed and treated like a criminal. Those of us who travel with guns will have to continue to develop imaginative workarounds while Nazis like this woman operate unrestrained.

2 July 2003

City Stick!

I've been interested in walking canes, as a cane is something one can openly have with him in just about any circumstance, without garnering the attention of others. A number of my friends and colleagues routinely fly with canes now and carry them as they travel.

To that end, I now have a copy of Cold Steel's City Stick. It is a stylish walking cane, but the handle is a metal knob, rather than a traditional arch. It is a formidable weapon! Like everything made by Cold Steel, it is extremely strong and well made.

I'll be flying with it this month and gauge the reaction. I need to get used to waking with a cane. Wonderful way to be armed!

2025 Update: I've since flown commercial hundreds of times, domestically and internationally, with my City Stick, all with no problems!

3. Women

21 Mar 2003

At a Women's Defensive Handgun Course in Los Angeles last weekend, we had two students of exceptionally small stature. Together, they probably didn't weigh more than 175 lbs, and both were slight of build, short, and had small hands.

Temperature was in the low fifty's, chilly for LA, and it was raining. The combination of circumstances caused both women to be unable to pull the trigger on their SIG P239s (9mm) unless the hammer was manually cocked first. Using the middle finger as a reinforcement for the trigger-pull helped a little, but accuracy was poor.

Both women were able to pull the trigger on a G19, with much better accuracy, but the fat, double-column Glock grip was clearly too big for either of them.

A Kahr 9mm (PM9) turned out to be the best compromise. Though not perfectly, both women could acquire an acceptable grip and simultaneously pull the trigger smoothly.

Happy to say that both women made up in resolve what they lacked in size and strength. Both made a good show.

Lesson: None of the foregoing would have come to light had we been shooting in warm temperatures and dry, sunny weather. Emergency equipment has to be suitable for all foreseeable circumstances. Icy determination and/or alternate shooting techniques (such as using two fingers to press the trigger or even manually cocking the hammer) may have to make up the difference when adverse circumstances make doing it the "regular way" impossible. Instructors must be sensitive to the foregoing.

The goal of both the student and the instructor is the same: *the improvement of the student*. Instructors must inspire as well as merely showing the way.

"I'd rather see a sermon than hear one any day.
I'd rather you walk with me than merely tell the way
The eye is a better student, and more willing, than the ear
I find your counsel confusing, but your example is always clear!"

8 Dec 2003

From a female friend in SA:

"Just to let you know what happened to me on the weekend. I was robbed, sitting in my car, window down (it's summer here), waiting for the light to change. This is in downtown Capetown.

I was on the phone, talking to my husband. A person came to my window distributing pamphlets. I mumbled that I wasn't interested. Next thing, I see a person (same one? I don't know!) at my window, bending down and grabbing the gold chain from around my neck. I grabbed hold of it with my left hand and tried to yank it back. Last thing I saw was my chain going out the window!

I yelled that I am being robbed, but no one paid much attention. I then went to the Waterfront, where I approached a security guard, who was helpful. He took me to their headquarters, took a makeshift statement, and called the police. An hour later, Sgt Adams of the Table View Police arrived, casually took my statement and gave me the case number, so I could make an insurance claim. There was no attempt to actually investigate the incident, nor identify and arrest the robber.

I had to confront to fact that, until now, I've been living in a dream world. I am responsible for my own safety? I always thought the police did that!

What have I learned? That I have been foolishly naive. I need to make significant changes in my routine, indeed in my entire personal philosophy. I really am 'on my own.' Aren't I?"

Comment: My friend learned an important lesson cheaply. All she lost was a piece of jewelry. She is lucky to be alive!

Yes, we are all "on our own!"

17 Dec 2003

Comments to his own friends on my last quip, from a friend and colleagues in SA:

"Have a look at this! Does your wife's kit measure up? If your lady does not carry all the time, shame on you. She needs a good pistol, good accessories, and good training.

Women are extremely exposed here in South Africa. Those who do not carry regularly are all raped and/or murdered, sooner or later. It happens hundreds of times every day here.

Don't be so foolishly naive as to think this doesn't apply to you. In my trade I see people daily with 'sad stories.'

She needs true freedom. Only armed and trained women can be genuinely free."

20 Dec 2003

On fashion, from a female LEO and colleague:

"John, I can remain silent no longer! Fashion doesn't dictate readiness. Readiness dictates fashion.

As you well know, for us, the ability to respond with deadly force is a continuous requirement, not an 'option.' I have to be heavily armed all the time. My 'work dress' ranges from jeans, T-shirt, and a cover garment, to business suits for staff meetings, to dresses for court appearances. For me, concealment is critical, because I don't have the option of leaving my gun(s) behind.

No circumstance will preclude me from carrying! Choose the handgun, then the holster, access the circumstances, and lastly select the clothing.

Make it work!"

Comment: I dare not!

4. Gun Handling

6 Jan 2003

I just had a conversation with a friend on active duty who just returned from the Mideast. He is a competent gunman and one of my students.

My worst fears were confirmed. Gun handling skills and philosophy haven't improved one bit since I was in Vietnam thirty-five years ago! Soldiers and Marines are still afraid of carrying loaded guns, even in areas of active fighting. They doubt their own ability, and they think they need permission to have their guns is a state of readiness commensurate with the circumstances. They are afraid to make any gun-related decisions for themselves.

Commanders still think empty guns are safer than loaded ones. So, they want everyone to unload their rifles and pistols before coming into certain areas (what are they afraid of?). "Clearing barrels" are provided for this purpose. Not only does this pointless practice waste valuable time, it seems that there have been so many UDs during the procedure that personnel have now been threatened with prosecution if the UDs continue. The "solution" is to lynch people for not exercising the training they've never had!

Commanders obviously know that the small-arms training troopers have received is so poor and irrelevant that they can't be trusted with guns. It is the same reason National Guard Troopers patrolled airports with empty guns.

I had been told the situation had improved since Vietnam. In reality, it is worse. We don't have professional gunmen. We have scared kids who haven't been trained and scared commanders who are afraid of their own men.

Guns & Warriors: DTI Quips – Volume 2

28 Jan 2003

I just returned from Memphis, TN where I did a program at Tom Givens' wonderful indoor range. Tom is a well-known trainer and works with many local police departments.

He indicated that it is not uncommon for officers there to carry autoloading pistols with the chamber empty. When they step to the line the draw, chamber a round, then reholster. Then they signal that they are "ready."

We can thank our "friends" in the movie industry for this, just as we can thank them for making movie after movie where the hero runs around with a pistol in his hand and his finger continuously wrapped around the trigger.

It seems, in all the action thrillers, the heroes are constantly making their pistols click and clack. It must add to the drama for the benefit of the people who watch this sewage. When Tom asks officers why they are carrying pistols with no round in the chamber, they sheepishly answer that they "saw it in the movies."

Tom stays busy!

On the same theme, I just received a message from a friend in the theater of operations in the Mideast. He reports that, even in forward areas, soldiers and civilian contractors alike are still carrying empty guns around, because they are ordered to do so. "Clearing barrels" are everywhere.

Fortunately, my friend pays no attention to such stupid rules, and neither do his men.

"Any fool can make a rule, and every fool will follow."

29 Jan 2003

More from a friend and student in the middle of things overseas:

"The silliness continues: Every deployed Battalion out here has its own

Gun Handling

'kabal' (depression surrounded by earthen berms). Each and every one, of course, has its own 'clearing barrels' at the gate.

When traveling between these kabals (most only a few kilometers apart), one is permitted to have a magazine inserted in his weapon (but no round chambered, of course, as that would be far too dangerous), but, upon entering a kabal, one is required to sterilize completely. As you might imagine, traveling from kabal to kabal (as I must do daily) is so absurdly frustrating and time consuming that we laugh about it. We are supposed to stop and exit our vehicle in order to clear our weapon(s) at each kabal. The time wasted with this idiocy is substantial. To add insult to injury, there is no place for us to 'load' (insert magazines) on the way out. That is apparently unimportant.

All of this because our 'leaders' unconditionally distrust warriors with loaded weapons. They are afraid of guns, and they are afraid of us. Knowing the poor training that most military personnel receive, I understand their fear, but instead of providing proper training, they order everyone to carry an unloaded weapon or no weapon at all. The 'cannon-fodder mentality' is alive and well over here.

'Enough of this nonsense already,' I said to myself. I carry concealed under my cammies in a Blade-Tech KYDEX holster (which I purchased myself). Most gate guards just assume I don't have a pistol and thus give me puzzled looks, but wave me through anyway. KYDEX is the best product for this environment.

I met our main body yesterday. They all arrived from the airport in (of course) condition three (magazine inserted; chamber empty). The first thing our S-4 (a VMI graduate no less) asked me, 'Sir, where is the armory? These Marines have loaded weapons, and we need to turn them in before they have a negligent discharge.'

I looked at him indignantly and replied, 'I'm sorry, lieutenant. I've obviously been misinformed. I was told that you folks were men of honor, men of integrity, fearless warriors, ready and eager to defend our country.

I see instead that you are pitiable and fearful worms, afraid of your own guns!' Not a word was said in reply, but they all slept with their weapons last night, and there were no NDs. Imagine that!

Anyway, I wish officers who were afraid of Marines carrying loaded weapons would find another line of work, preferably in the UK. I, for one, wear my (loaded) pistol constantly. After only one day of being screwed with, my captains all do likewise. We're slowly spreading the sunshine here!"

Comment: Good show, Colonel!

31 Jan 2003

Follow-up from a LEO friend on the West Coast:

"In 1994, I attended an ASLET (American Society for Law Enforcement Trainers) Training Seminar in Washington, DC. On one of the seminar evenings, all members were invited to attend a memorial service at the Law Enforcement National Monument. Over five hundred sworn officers attended the service.

We were also gently 'reminded' that, since we were in Washington, DC, none of us were 'allowed' to carry firearms. During the service, an officer behind me said, 'Look at that, five hundred unarmed police officers.' Elbowing the butt of my concealed Glock, I replied, 'I don't see them, but I do see five hundred felons.'"

Comment: Foolishness like this is obviously not confined to the Washington, DC. While it all gets sorted out, the art of discrete concealment is critically important, and we all need to get creative on this point. There will always be someone who thinks you shouldn't have a gun, but, so long as it stays (1) out of sight and (2) out of conversation, the whole issue is moot.

As my friend pointed out, discrete concealment is an important skill even for soldiers at the forward edge of the battle area. Better the weapon be out of sight than have some West Point pretty boy throw a hissy fit upon seeing it!

24 Feb 2003

From a friend and student stationed in country:

"While standing near a dreaded 'clearing barrel,' our Group SgtMaj casually pulls his pistol from his shoulder holster and, with his finger firmly in contact with the trigger, points it directly at the Marine standing in front of him.

Standing behind him, I saw what was happening and quickly grabbed his trigger finger pulling it out of contact with the trigger and placing it on the pistol's frame, while immediately elevating the muzzle. I said, 'SgtMaj, PLEASE keep your trigger finger off the trigger and PLEASE keep your pistol pointed in a safe direction, not at the liver of the fellow Marine in front of you.' He was slightly embarrassed, but quickly rebutted 'Colonel, I have been shooting firearms all my life.' I responded, 'I'm sure, but you don't seem to have learned the first thing about correct gun handling in all those years.' Needless to say, my captains are refusing to stand anywhere near clearing barrels when this guy is around, as incidents like this are not uncommon, even among officers and senior NCOs.

Unhappily, this 'I already know more than you can teach me' attitude is widespread amongst the 'seniors' here. As always, we have an uphill battle."

Comment: Will we ever learn? Must we go through this all over again at the beginning of every war?

28 Apr 2003

From a friend who shoots a lot of competitive rifle:

"I own a Garand I acquired ten years ago from the CMP. I was practicing with it for a match yesterday when the case of the fifth round separated in the middle. The forward half remained in the chamber, and the ejected half was found fourteen yards in front of the firing line. I got a good blast of hot gas on my face and glasses but was otherwise uninjured. Ammunition was new, Federal American Eagle 150gr FMJ. Firing had been slow, because I was working on my standing position. The rifle appears to be undamaged.

I have read about head separations with reloads, but, with new ammunition, it isn't supposed to happen."

Comment: No, it isn't! The lesson here is that we all need to discipline ourselves to wear glasses and a baseball cap any time we're shooting any kind of firearm. If it's "unusual," it will happen to you sooner or later, and always when you least expect it!

16 June 2003

From a friend who owns a gun shop in the south:

"A shooter at a big local match, while loading a handgun using the 'front slide serrations' on his pistol, allowed several of his left-hand fingers forward of the muzzle (an error nearly impossible to avoid when using this specious gun handling technique). Somewhere in there, his strong-side index finger made its way inside the trigger guard, and the pistol discharged. The shooter was astonished, as he was obviously unaware of where his fingers (on both hands) were. His left index finger was shot off, and he then pressed the trigger a second time in an apparent startle response and shot off the middle finger on the same hand!

The damage is not life threatening, but he now has a permanent, disabling injury."

Lesson: I don't know how often this has to happen! Grabbing the slide of any autoloading pistol ahead of the ejection port is not only a poor tactical procedure (because the ejection port is thus usually blocked), but also a veritable invitation to a shooting injury to the shooter's support-side hand, as we see from the foregoing.

On any autoloading pistol, placing slide serrations ahead of the ejection port is a senseless and dangerous design flaw, in my opinion.

13 Nov 2003

Winter carry option, from a friend who lives where there is genuine winter:

"I know it's potential is limited, but during the coldest times of the year, I routinely carry a J-frame snubby in my outermost coat pocket. In cold weather, there is surely nothing unusual or suspicious about someone walking about with hands in his coat pockets.

As is always the case when a carry pistol is not attached in some way to the body, it must be appropriately secured when the outer coat is removed. Generally, I discreetly take the revolver out of the coat pocket and put in into my right, front pants pocket, reversing the process when I put my coat back on.

Needless to say, the little revolver will always be supplemented with a more substantial fighting handgun concealed in a belt holster, but, as I walk about, I find five rounds of Cor-Bon 38 special at my fingertips and instantly ready to shoot through my coat very comforting indeed!"

Comment: S&W's lightweight (scandium) snubbies are particularly well suited to this carry option. The entire hammer arc is contained within the pistol itself, so there is no chance clothing will get in the way of the hammer path. And, the pistol is so light, carrying it in any fashion is about as convenient and comfortable as the practice gets.

2025 Update: Cor-Bon has changed owners twice since this was written. Today, Defiant Munitions represents a good choice.

5. Techniques

12 Mar 2003

I've been teaching a new gun/blade technique that I learned from some experienced defensive blade instructors in Africa, Mark Human and Kelee Arrowsmith, from the local AMOK system. It integrates well with what we're doing already and is effective.

We position the blade in the weak-side hand much the same as we position the flashlight in the Harries Technique. The blade becomes, in effect, a bayonet, making it difficult for anyone to grab the gun without getting cut.

The user can thrust forward with the blade in conjunction with a verbal warning in order to keep potential gun-snatchers at bay. The blade can also be use in a similar display to keep people from approaching from the sides.

When your blade makes contact with someone who gets too close, the verbal command that works best is, "You're cut! You need to go to a hospital." That gets his mind off of fighting and onto his own situation, turning his thinking inward, rather than outward. A blade that is dead sharp (like those from Cold Steel, Hoffner's) will cut grievously with little effort and often little sensation on the part of the recipient. However copious bleeding is easily seen once attention is called to it.

I have become convinced that this is a good technique that we should all have in our repertoire, as we all normally carry both a pistol and a knife. It (1) enables one to remain in control, both of the situation and of his pistol, (2) providing a substantial deterrent to close approach, and (3) effectively preventing anyone from executing a successful gun grab.

Knifes are becoming more useful all the time!

3 Aug 2003

Fox OC works! This from a friend with the Capetown Traffic Police:

"Saturday, I was manning a roadblock during a big soccer game at our local stadium. The post was set up at one of the main pedestrian exit points. I was standing there in uniform after the game while hundreds of exiting fans streamed by. I had a can of the Fox OC in my weak hand and my Streamlight tactical torch (flashlight) in the other

I locked on to a large male approaching me. He was obviously drunk, and he was holding a glass bottle in one hand. As he approached, he rushed toward me, raising the bottle, and shouting, 'Ek gaan jou moer,' (loosely translated, 'I'm going to make a mess of your face).

I immediately turned on my Streamlight and directed the beam into his face, as I moved sideways, off the line of force. He instantly became disoriented and had no clue as to where I had gone. Next, he got a face full of Fox!

One step backwards, and down he went, dropping his bottle, and 'flopping like a fish,' as you might say. He was no threat to anyone from that instant forward. The group he was with scooped him up, and off they all ran. None of the others seemed interested in sampling their friend's plight! It all transpired so fast that my partner wasn't even aware of what had happened. A few passersby started coughing, but most had no idea of the drama that just took place in front of them.

Although Fox OC has about the same 'percentage' that the crap they issue to us, its knockdown power and overall effect are vastly superior, which is why, I'm sure, they don't issue it to us. However, I'll never be without it!"

Comment: A powerful flashlight and a bottle of Fox OC makes a formidable, non-lethal combination. The alliance surely worked in this case!

2025 Update: Fox has since been absorbed by Sabre. I currently recommend Sabre's OC.

Techniques

6 Dec 2003

From a friend and student:

"The recent abduction/disappearance of a North Dakota college student and the murder of a federal prosecutor illustrates, once again, the three deadly omissions:

(1) Inattentiveness
(2) Unpreparedness
(3) Defenselessness

Inattentiveness: If you are routinely unaware, you can take steps to make yourself even more so by talking on a cell phone as you walk in dark parking lots. How common, and how foolish!

Unpreparedness: Perceiving, in time, a set of circumstances that are suspicious is always the first step. Failing to have a plan makes it all for naught! In the absence of a plan of action, most people, when confronted with criminal violence, panic and lapse into a state of mental paralysis. While they dither, the predator makes his move,

Defenselessness: When a physical fight is unavoidable, you need to be able to fight effectively. The unwillingness or inability to fight effectively is a virtual death sentence when criminal violence comes your way. If you allow yourself to be tied up, taken by force to a remote location, etc, your chances of survival are essentially zero. You need to be prepared, equipped, and willing to 'make your move,' when making your move still has some likelihood of success.

All of us, even those of us who don't routinely carry concealed guns, can have OC spray with them most of the time. All of us, in most places, can carry a serious blade. Not having such easily carried defensive instrumentalities on your person all the time is silly.

Most people's ONLY 'defense' against criminal attack is the perpetual hope they will not be selected as prey by a predator. Deselection strategies are surely an important part of any plan, but effective fighting is a part of every legitimate plan too, and most of us do not think about such things nearly enough. Like discussions of smoking, drugs and AIDS, such topics are uncomfortable. But 'thinking about the unthinkable' is a lot less uncomfortable than what these two most recent victims went through."

6. Tactics

19 Mar 2003

Layers of response:

Years ago, Jeff Cooper delineated the "Color Code" and the "Principles of Personal Defense" in an effort to provide us with a logical model for one's thinking on the subject of mental preparedness. I'd like now to go to the next step and apply the same logic to the issue of personal appearance and demeanor, as we all agree that, in the domestic defensive environment, avoiding a fight is always preferable to "winning" one.

Layer One: Nonattendance. The best way to handle any potentially injurious encounter is: Don't be there. Arrange to be somewhere else. Don't go to stupid places. Don't associate with stupid people. Don't do stupid things. This is the advice I give to all students of defensive firearms. "Winning" a gunfight, or any other potentially injurious encounter, is financially and emotionally burdensome. The aftermath will become your full-time job for weeks or months afterward, and you will quickly grow weary of writing checks to lawyer(s). It is, of course, better than being dead or suffering a permanently disfiguring or disabling injury, but the "penalty" for successfully fighting for your life is still formidable.

Crowds of any kind, particularly those with a political or religious agenda, such as political rallies, demonstrations, picket lines, energy-charged sporting events, etc are good examples of "stupid places." Any crowd with a high collective energy level harbors potential catastrophe. To a lesser degree, bank buildings, hospital emergency rooms, airports, government buildings, and bars (particularly crowded ones) fall into the same category. All should be avoided. When they can't be avoided, we should make it a practice to spend only the minimum time necessary there and then quickly get out. I'm in bed by 10:00pm. No late-night excursions for me!

"A superior gunman is best defined as one who uses his superior judgment in order to keep himself out of situations that would require a demonstration of

his superior skills."

Layer Two: Functional invisibility. We all need to practice to art of "being invisible." It is in our best interest to go our way unnoticed, both by potential predators and by the criminal justice system alike. We call it "The Stealth Existence"

Whenever I travel, particularly to foreign countries, I endeavor to be the one that no one notices; no one recalls; no one remembers. I silently slip under the radar, leaving no trace, a nameless, faceless tourist. When in any public place, I try to be clean and well groomed, but I never wear bright colors, nor any kind of flashy jewelry, nor or anything shiny. I smile a lot, but talk softly and as little as possible. I try to be polite, pleasant, boring, uninteresting. As we say in the law enforcement business, "Courteous to everyone. Friendly to no one."

Loud talking, bright colors, Rolex watches, etc. will consistently accumulate unwanted attention. On the other end of the spectrum, tattoos, poor grooming, loud/offensive language, a slovenly appearance, etc. will also garner unwelcome notice.

Layer Three: Deselection. Any successful predator has the ability to quickly screen potential victims, focusing in on the ones who look as if they will make good victims and rejecting those who either (1) look too strong/alert for expedient victimization or (2) don't conveniently fall into any particular category.

When invisibility fails, we endeavor to be consistently deselected for victimization. We do this by making it a habit to appear alert, uninviting, self-confident, and strong. At the same time, we never loiter, nor appear indecisive. We are always in motion.

"Weakness perceived is weakness exploited!"

Layer Four: Disengagement: Our best interests are not served by any kind of engagement with potential predators. Successful disengagement involves posturing, bearing, verbalizations, and movement. It is in our best interest to disengage at the lowest reasonable force level, but we must simultaneously be prepared to instantly respond to unlawful force with superior force.

Potential predators, as they attempt verbal engagement, should be politely dismissed. Bearing and eye contact should always project strength and confidence. Our verbalizing should never be demeaning, nor threatening, lest we be accused by a prosecutor of starting, or aggravating, the encounter.

We should continuously be moving off the "line of force." We should be observant in every direction, giving potential predator duos and trios the distinct impression that they will not be able to sneak-up on us.

When predators are confused, they are unable to focus sufficiently to carry off their victimization. Therefore, never let a potential predator seize the agenda. Don't answer his questions. Don't engage in any kind of conversation, and don't stay in any one place very long.

Disengagement, separation, and exit are our immediate goals when we have been selected or are being seriously evaluated by predators. However, if there is to be a fight, the best one is a short one. When a predator menaces me with a gun or knife, I know that, before it is all over, there is a good chance that I will be shot or cut. However, within that prison of circumstance, I also know that the faster I can end the fight, the less hurt I'm going to get! When there must be a fight, I must explode into action, moving smoothly and quickly, in an effort to confuse and overwhelm my opponent before he has a chance to process all the information I'm throwing at him.

Ultimately, we must "have a plan." Potentially dangerous encounters must be thought about in advance. Decisions must be made. Skills must be practiced. Confusion, hesitation, and vacillation will always attract the attention of predators and simultaneously stimulate their predator behavior.

Guns & Warriors: DTI Quips – Volume 2

5 June 2003

2003 NTI, Harrisburg, PA

This year, I was only able to stay for my day of shooting/testing at the NTI. Then, Vicki and I had to drive back to MI for courses there, which is why this summary is late this time. My usual practice is to stay for the entire NTI week in Harrisburg, PA and attend the banquet on Saturday. Happily, both Vicki and I were able to participate in a panel discussion on Tuesday evening at the hotel.

Skip, Jim, and the crew did yeoman's work putting this year's event together. Most of the setup had to be done in the rain. In fact, it rained most of Tuesday when I shot the course. Everything, including all the electronics, worked, in spite of great quantities of mud, flowing water, and puddles (actually lakes!) everywhere. We were all a soaked, muddy mess at the end of the day.

Good show, once again!

As always, this year's event included Simmunition drills in ASTA Village. Everything else was live fire. Included were: a "strange weapon" drill in a darkened building, two "escorted" rescue drills in building mockups, two "standard (known)" tactical exercises, and an "all by yourself" rescue drill in the 360 building mockup.

The day was fatiguing, uncomfortable, and unsettling as always. The rain and mud made it all the more intensive and exhausting. This year's challenges were the most difficult and sophisticated yet. Most of what I encountered in the building mockups was unexpected, and I had to both think AND react my way through. It is the most mentally and emotionally draining activity in which I participate all year, with the possible exception of my big game hunting expeditions in Africa. The two activities are surely on the same level.

This year, I used my G32 (357SIG), Comtec IWB holster, Rusty Sherrick shoulder holster, and my S&W M340PD for backup. I carried my usual three knives (all by Cold Steel), a Ti-Lite for speed, a Culloden neck knife for stealth, and a Vaquero Grande for serious fighting. All was concealed under my CCW Clothiers vest.

In my G32 I shot Cor-Bon 115grHP. In my snubby, I shot Cor-Bon 125grHP. Nary a hiccup from either gun. All my Glocks are stock, except for factory night sights and a (standard) NY trigger on a five-pound connector. Most other after-market modifications are "not a good idea," in my opinion. My S&W scandium snubby is completely stock.

Here are this year's lessons:

ASTA Village:

When you have multiple, potential opponents, stack them.

In one drill, I was at a gas station when I was approached by a panhandler. Assuming the interview stance, I snapped my head around and noticed a second sleazy character behind me. By the time they both drew guns, I was between them. I should have stacked them when I had the chance, keeping both in front of me. I failed to take advantage of the opportunity when it presented itself. As things turned out, I was forced to stack them during the gunfight.

Had I not made it a practice to look all around, I never would have seen the second assailant. This was the sad fate of many of my fellow participants.

Indecisive movement is scant improvement over standing still.

In this same drill, I should have moved off the line of force boldly, as soon as I smelled a mugging in the making. Instead, I dithered as I processed all the incoming information. I was moving somewhat, but my insubstantial motion communicated indecision and confusion, instead of resolve and strength.

Verbal commands and addresses need to be practiced.

If they are not, the come out garbled and indecipherable. Tape loops must be rehearsed until they flow smoothly.

Cunning criminals can easily separate a member of a group from the main congregation.

My wife and I were walking side by side. Before I knew it, two mugging suspects had grabbed her and were carrying her away, kicking and screaming. By the time I realized I had no choice but to shoot one in the back, as he (and she) were already three meters away.

Leave when things start going in the toilet.

In another challenge, I was filling out a form when a person came into the same office. His conversation with the clerk started normally enough, but rapidly grew belligerent. I wanted to complete the form, but decided I needed to exit. Sometimes, it takes a while to reaffirm one's priorities. I took too long!

In a crisis, confusion rules. Emergencies need to be discussed, even rehearsed, in advance.

This year, after I thought the ASTA Village exercise was over, we were all asked to step in to a conference room to discuss the segment. Suddenly, a bomb went off! It was loud. Walls buckled, and the ceiling caved in. The room was filled with smoke. Gunshots and yelling were heard. I, for one, was taken completely by surprise. It was obvious to me that I had not thought enough about such an incident!

Don't panic. Think your way through.

One of the live-fire segments involved the necessity to use an unfamiliar weapon (a British SMLE Enfield in 303Br) which I had to use in a darkened room. My predictable reaction to mission overload is always exasperation and dread. The mud and the rain also did nothing to improve my focus! I have to force myself to take a deep breath, regain control, size up the situation, make a plan, and move out boldly.

Carry pistol and backup pistol need to be used in harmony.

Tactics

In several of the elaborate, live-fire challenges, after the second magazine in my G32 ran low, I consciously holstered it with several rounds left as I drew my backup pistol. In effect, I manufactured another backup. It was surely reassuring to me that, when I ran dry, I could immediately default to another gun.

The knife/pistol technique works well when one is surrounded by people who might want to disarm him.

I used my Cold Steel Vaquero Grande to dissuade potential gun grabbers in several of the problems. Upon seeing my blade and gun in such close proximity, few had any interest in attempting a disarm

When a building fills with smoke, I was astonished at how disoriented I became.

I couldn't remember where I had been, and I couldn't find the exit. Once again, I had to calm myself down and make a plan.

Look up and down, as well as level.

I missed several important clues, because I failed to look up and down. In any tactical circumstance, one has to be moving and looking continuously.

Move laterally upon seeing a threat.

I am always tempted to stay in place and fire the instant I see a threat, but I now force myself to move off the line of force as I'm presenting my pistol.

One-handed shooting is a much more important skill than most realize.

Again, this year much of my shooting was one handed, not by choice. My left hand was constantly occupied with bloody bodies, opening doors (and holding them open), and keeping myself from stumbling.

Standing targets need to be zippered.

The zipper technique (starting at the navel and working one's way up the body midline and into the thoracic triangle) routinely prevented me from getting my front sight too high too fast.

Again, all serious gunmen should try to get to the NTI. We all leave our egos at the door and lapse into "student mode." I wouldn't miss it!

2025 Update: The NTI (National Tactical Invitational) is no longer held, but Tom Givens' Tactical Conference (Tac-Con) is currently a yearly event and highly recommended!

7 June 2003

Fighting from within cars:

A federal agent and great friend of mine and of the NTI presented a thoroughgoing lecture on fighting from within cars during this year's event. Important points:

A moving car, going as slow as fifteen miles per hour, is seldom penetrated by pistol bullets, even rifle bullets. Resistance to penetration is even greater at higher speeds. A stopped car is penetrated much more often. The lesson is clear: if you're in a car and the fight starts, stay in the car, speed up, get everyone down, and get out of there.

Unless the attackers have RPGs, staying in the car and leaving at high speed nearly always makes more sense than exiting the car and fighting/fleeing on foot. A hit from an RPG will cause casualties on the inside of a car and probably disable it, but even a skilled RPG crew will have great difficulty hitting a rapidly accelerating car. As noted above, small arms fire at moving cars mostly fails to hit in the first place, and even the rounds that do hit seldom penetrate to the interior.

Molotov cocktails are largely ineffective against moving cars. If one strikes your car, just drive away. It will burn out in seconds and so little damage.

Keep the interior of your car clean. Trash inside a vehicle can become harmful missiles during an explosion or high-speed crash.

Seat belts are a two-edged sword. Wearing them restricts movement and makes getting into an effective firing position difficult. However, wearing them also makes it likely that one will survive a crash and remain conscious long enough to exit the vehicle and flee to safety. Wearing them is usually a good idea. They should only be taken off when things get desperate.

Crack the windows several inches when things start going in the toilet. A cracked window is easily broken. Rolled all the way up, car windows are nearly impossible to break from the inside.

When you're being followed by another vehicle, crack the driver's window and spray OC out the opening. The slipstream will deliver it directly to the pursuing vehicle. Even when their windows are rolled up, the OC will be sucked in to their vehicle and "encourage" them to find something else to do!

When firing at pursuing vehicles, shoot into the radiator. It's a big target, and once a radiator leak is created, the pursuing vehicle will quickly overheat and have to stop.

The best weapons for fighting from a car are pistols and short-barreled rifles. Long-barreled rifles and shotguns are unwieldy and difficult to maneuver inside a car. A good "urban rifle" is just the ticket.

1 Nov 2003

From a friend in the LA area:

"Just saw on local TV newscast footage from LA of man shooting at an attorney with a handgun (looks like a five-shot snubby revolver) at pointblank range. The victim, wearing suit and tie, was struck at least once, but the hit(s) appear to be peripheral and not life-threatening.

The attorney, while being shot at, continued to resist vigorously, dodging behind a tree and using it as cover, with great skill, while his assailant danced on the other side trying for another shot. The assailant quickly ran out of ammunition. Instead of reloading and finishing what he had started, he casually walked away and was immediately subdued by police. The victim walked away too, fully cognizant and displaying little discomfort!

Not often do we see so clearly evidence that pistol rounds are frequently ineffective at stopping people from acting."

Lessons:

(1) Keep fighting as long as you are conscious. Impacts from most pistol bullets are not instantly incapacitating, unless you've convinced yourself that they should be.

(2) On the other side of the ledger, pistol rounds can't be depended upon to make the action stop. Multiple impacts, even reloading, may be necessary before the fight ends.

(3) Adept use of cover is a critical skill. Cover is a life saver, even at close range

(4) Don't go to LA without a gun!

2 Oct 2003

Fox Incident in SA:

My friend recently underwent back surgery and is currently in a brace and unable to sit down:

"My wife drove me to our local shopping mall, so that we could take a short walk. Car transportation for me consists of an awkward process of arranging my frame into a suitable configuration, so that I can plop into the passenger seat. The seat is flat, as I cannot sit. Getting out is even more of an expedition.

On arrival at the mall parking lot, my wife parked the car and got out to come over to my side to help me. I unlocked my door and opened it partially. Next thing, the door was violently yanked open, and I was confronted by an aggressive, belligerent drunk, who was shouting, cursing, and demanding money. No hesitation on my part! He instantly got a squirt of OC. My aim was off. The main stream went wide. However, the little he did get was more than adequate! He lurched backward in astonishment. The last I saw of him, his face was buried in his hands, and he was coughing, gagging, and floundering aimlessly on one foot and one knee. My wife and I, of course, left immediately and parked on the other side of the mall. Never saw the guy again."

Lesson: My friend had a plan! The drunk did too, but the last thing he expected was an immediate and precise counterattack. In the end, a violent encounter was avoided, and my friends are okay. Good show!

OC works most of the time! Highly recommended. I carry it all the time.

6 Nov 2003

OC Advice from an SA Patrolman of many Years:

"Tactical plans always have two parts: The first is to actually have a plan or 'blueprint of action.' The second is always having a means of implementing your plan. This second part is where lots of people see their plan fail. Without it, your 'plan' remains just a theory.

To this end I have a simple routine that I always follow when performing traffic stops. I make sure both my pistol and my OC are ready for instant use upon exiting my vehicle. This procedure has spilled into my private life. I always make sure that my OC is handy whenever I bring my vehicle to a stop, in other words, whenever I drive my vehicle. I recommend 'practical' OC, like Fox (now Sabre), that is of a size that it is easily concealed in your hand with a nozzle that can be 'aimed' by feel. I have seen more than one officer accidentally spray himself, because he had no 'felt aim' on his OC bottle.

I also wear a sleeveless jacket. These normally have big pockets. I designate one of them my 'OC pocket.' Nothing goes into that pocket except my OC. Surprising how readily your OC bottle comes to hand (in the correct orientation) when you make alertness and readiness a habit!"

23 Nov 2003

Length of domestic gunfights:

At a recent gathering of police trainers and range officers, a lecturer posed this question: "How many of you guys have ever arrested a person and subsequently, during a personal search, discovered and removed a concealed gun from them? For that matter, how many of you have ever seized a gun during a pat down?" The response was immediate and 100%. Everyone there had done it multiple times.

He acknowledged the response and then asked: "Okay, how many of you, upon discovering a gun on an arrestee, have ever also discovered spare ammunition in the form of fully charged spare magazines, speed loaders, or a second gun? The response was just as dramatic. None ever had!

This pattern appears to be common among at least criminals in this country. Violent domestic criminals, armed robbery suspects and mugging suspects, when they do carry a gun, almost never carry any more ammunition than is in the gun itself. They are not notorious for long-range planning!

The implication for us on the other side is: The most important part of the gunfight is the first five seconds! After that, most armed criminals will have expended all rounds they have and will have no ability to reload. Having the tactical knowledge and skill (movement, use of cover, tactical planning, shooting and gun handling acumen) and equipment to survive the initial burst of violence is critical. After that, victory is virtually assured (not that we should ever relax). We need to be able to literally outlast any opponent.

Part 2: Case Studies & Incidents

7. Accidental Discharges (ADs)

14 Oct 2002

We just completed and Advanced Defensive Handgun Course in the Midwest. We had a number of students join us from a large, local metro PD. This PD permits its officers to use virtually all handguns, except Glocks. The instructor staff (whom I had in the class) has been pushing for Glock acceptance, but to no avail.

The official cover story is that this department doesn't like "striker-fired guns." The real reason is that one of the department's "commanders" had an UD with a Glock as he was trying to put it back into its shipping container. Seems he pressed the trigger having neglected to unload it first! Being a product of our times, he, of course, assumed no personal responsibility for his stupidity. He, instead, blamed the gun, and Glocks have been blackballed ever since. The training staff is hoping he retires soon!

5 Mar 2003

Locker room UD:

"Our department (suburban, Chicago) had an AD in our locker room recently. Gun involved was a G22 with 180gr HP (Gold Dot).

The single round went through the sheet metal locker shelf and back panel, three dry-walled walls (total of six layers of drywall), and across two hallways. It continued through an office (narrowly missing an officer sitting therein), ricocheted off the office floor (concrete) and then went up through an open doorway and penetrated a hallway ceiling tile, where it finally came to rest. It traveled a total of ten meters. There were a few anxious moments but no injuries. The bullet's hollow point cavity was plugged with drywall, and the only deformation was from its impact into the concrete floor.

The involved officer was not officially disciplined, but he did get a stern talking-to by the chief and others. No report was made. Damage was quietly repaired. As far as the department is concerned, it never happened."

Comment: The above is pretty standard when there is a police UD that does not involve personal injury. True statistics on police UDs can thus only be estimated.

7 Mar 2003

Good comments on police UDs from a friend and well-know trainer:

"From my perspective, it is good that this department did not overreact, by instituting 'clearing barrels,' 'gun-free areas,' etc. Many state police academies are now 'weapons free zones' because of the concern over UDs. One facility (where I do training) now mandates that no firearms are permitted in any building, all because one cop had an UD in the bathroom. It only happened once.

The fact is, people are human, and when we have large numbers of armed individuals, the price we pay is that accidents happen. Through training, we minimize them, but they can never be eliminated completely. We have vehicle accidents too, but we don't ban vehicles immediately after the first one. We consider them 'lessons learned' and then press forward.

In the face of elevated terrorist threats, any concentration of cops who are known to have been disarmed is an enormously desirable target. Such attacks have already happened with unarmed Marines in overseas facilities.

'Gun-free' zones are actually 'criminal empowerment' zones, and we are going to pay a dreadful price when we allow them to become a trend."

8. Non-Police Gun Owner Incidents

10 Jan 2003

From a friend in Atlanta:

"Several days ago a seventy-year-old shopkeeper was sitting in the back room of his package-goods store down here watching TV when he noticed three sleazeballs enter his store.

He calmly picked up his Remington 870 (charged with 00 Bk), and, utilizing his walker, hobbled out to the counter with the pump held low along one leg of the walker and below counter level.

The three nervously announced a holdup and pointed a pistol at the shopkeeper. Surprise! The shopkeeper had been through this drill before. He calmly raised the shotgun, chambered a round, and began firing at the astonished robbery trio. As a result, two robbers were DRT and the third is critical (at this writing, still hanging onto life by a thread). The robbers never got off a shot.

Police arrived and arrested the wounded suspect (who didn't get far). They also arrested the robbers' stupid bimbo/girlfriend, who was still sitting in the getaway car twenty minutes later, apparently not paying attention.

Last year, this same shopkeeper was held up by another armed suspect. In that incident, the robber shot the shopkeeper without ever saying a word. The robber was also shot by the shopkeeper after the shopkeeper had been wounded. That robber was DRT also. Quoth the shopkeeper of his wound, 'It was through-and-through; didn't hurt'. He obviously saw no reason to give up the fight!

He has no plan to retire, nor to sell his business"

Lesson: You're only beaten when you allow yourself to believe it. Criminals are wimps. An explosive counterattack will always take them by surprise. When an elderly and crippled shopkeeper can be victorious in a situation like this, what are we to think of healthy cowards who fearfully give up the fight before it even starts?

29 Jan 2003

Good news from an attorney friend:

"Two days ago, the Judge in West Palm Beach, FL in the Raven pistol case, ruling on post-trial motions, issued a directed verdict overturning the entire judgment against Valor Corp, the pistol's distributor.

You will recall this is the case where 13-year-old Nathaniel Brazil, upon being suspended from school, went home, retrieved a 25ACP Raven pistol he had stolen several days earlier from his 'grandfather' (actually no relation), returned to school with the gun, and shot his teacher in the face, killing him. Brazil was tried as an adult and convicted of murder. This present case is a civil suit, sponsored by Handgun Control Inc, against Valor Corp.

With the judgment against Valor now being overturned, the only remaining judgments are against the school district (which wasn't a party so isn't liable to pay the judgment) and the elderly 'grandfather' (judgment-proof by virtue of no significant assets) who owned the pistol and kept it unsecured where Nathaniel Brazil could take it.

The jury found that the pistol in question contained no design defect, no manufacturing defect, and was not 'negligently made' (whatever that means). In ruling on the motions, the judge agreed that the fourth issue, the only one on which the jury had found Valor liable (for not selling the pistol, in 1988, with a trigger guard lock) was improper under Florida law.

This is a tremendous victory for the good guys, as Handgun Control Inc was trying to prove the Raven pistol was 'inherently defective,' simply because it was small, inexpensive, of small caliber, and therefore had 'no legitimate purpose other than as a crime gun.' NO gun manufacturer provided trigger guard locks with their pistols in 1988, when the Raven was sold, and Valor wasn't even the manufacturer. The manufacturer, Raven Arms, has been out of business for years.

If allowed to stand, a decision against Valor in this case might have been usable to say that ANY seller of ANY gun, new or used, big or small, at least, as far back as 1988 if not further, would be liable for criminal misuse of the gun or for gun accidents."

Comment: You won't hear about any of the above on the mainstream media, as it is bad news for them and their left-wing agenda. When good news like this comes along, being the ideological prostitutes that they are, they simply pretend they didn't see it.

15 May 03

From a friend in Capetown, SA:

"I just got off the phone with an attorney friend. He was accosted today by two armed-robbery suspects. They barged in to his office, grabbed him, and threw him to the floor. He drew his G23 as he went down and fired as soon as he got his front sight on the nearest one.

Both suspects were hit once. One was then hit a second time. The nearest one was shot once in the face. He collapsed immediately but was still breathing when my friend last saw him. The other was hit twice in the trunk. He fled the office stumbling and bent over.

Police were called two hours ago from the office next door and still have not arrived.

My friend carries his G23 in a Fobus paddle. This one apparently worked fine. His G23 was loaded with Speer Gold Dots.

My friend is shaken but otherwise okay."

Lesson: When it's least expected, you're elected! This lawyer was armed and ready. Most importantly, he had thought all this through in advance. He had a plan. At the moment of truth, he did what he had to do to keep himself from getting hurt. Good show!

9. Police Incidents

24 Oct 2002

From an LEO friend in Wisconsin:

"… just after dark, the burglars returned and all were quickly arrested. The chief investigator then told me, 'I'm glad I was wearing my gun that day.'

I asked him what he meant. He indicated he had been processing worthless checks turned in by local merchants all day, and that he doesn't always 'arm up' when he figures to be working around the office.

He must have noticed the look of astonishment on my face, because he quickly added, 'I've been in investigations for four years now. Guess I'm losing touch with the street, huh?'

I nodded in agreement.

He has allowed himself to quit thinking of himself as a police officer and began thinking like and as a bureaucrat. This kind of thing is distressingly common around here."

Lesson: You're either in the Navy or you're not! "Cop" is not just an occupation. It is a way of life. "Police" is not what we do. "Police" is what we are. Grasseaters need not apply!

4 Apr 2003

Shooting incident in South Africa:

"Two of our city police officers were in pursuit of a suspected stolen vehicle last Thursday. The driver of the pursued vehicle lost control and crashed. Our officers stopped and approached the vehicle in order to assist the injured and were promptly fired upon from within the crashed vehicle. Both officers were hit in the lower extremities. Suspect(s) escaped. Neither officer suffered life-threatening injuries, but both were hospitalized."

Lesson: Sometimes we make too many assumptions! Injured suspects can still represent a deadly threat. Error on the side of caution!

30 July 2003

From a friend with the NJSP:

"Several of my colleagues and I have been sent to 'protect bridges' between NJ and NY. Our brass decided that one, maybe two, troopers per bridge is adequate, in spite of the fact that thousands of vehicles drive over every hour. Joining us are members of the NJ National Guard, usually two and a Humvee.

Keeping in mind what you said about the National Guard in airports, I made it a point to ask the members of one NG detail deployed at a bridge about the supply of ammunition they had for their M-16s. You guessed it! They had no ammunition; none on them; none in their Humvee. They didn't even have magazines, just empty rifles.

We asked them what they had been instructed to do if things went in the toilet. Their response was, 'Our commander told us that, if there is a riot or terrorist attack, we are to run back to our Humvee and lock ourselves in!'

Needless to say, we consider them to be nothing but excess baggage."

Comment: It is bad enough that we have cowards in the command structure of the NG who obviously look upon their guardsmen as completely and utterly expendable. The real crime is that they demand that soldier themselves become cowards too. "Contemptible" is not a strong enough word!

15 Aug 2003

Police here and there:

A friend and student from the UK recently attended a course here in Colorado. He greatly enjoys shooting and realizes the importance of firearms skills, but, of course, in the UK any kind of defensive shooting training is nearly impossible for anyone but the politically connected. So, he joins us in the mountains once a year.

He shared with the class an interesting observation: He said, in observing American police officers, he noticed their attitude is one of "service." They look upon themselves as "public servants." Conversely, in Europe (even Western Europe), police look upon themselves as "public regulators." They are told their job is to "modulate" the public and, most of all, provide protection for government officials.

Some might call it a subtitle, even inconsequential, difference, but, coming from his mouth, it caused us all to realize what a great country we live in. Government officials and employees looking upon themselves as "public servants," rather than "public regulators."

What a concept!

Not surprisingly, all tyrants and tyrannies see themselves as "benevolent," but, as Alexis de Tocqueville pointed out, the will of man is thus not shattered by "beneficent regulators," but it is slowly softened, bent, and, in the end, asphyxiated. Men are seldom forced by it to act, but they are constantly restrained from acting. Such a power may not destroy in the physical sense, but it prevents true existence. It prevents one from claiming his own magnificence. It may not exactly tyrannize, but it squeezes, enfeebles, extinguishes, and ultimately smothers a people, until they are reduced nothing better than a flock of timid and fearful sheep, of which the government is (of course) the "shepherd."

22 Aug 2003

More Rifle comments from a LEO friend in the Midwest:

"We purchased Ruger Mini-14s last year and put them in all the cars. They 'feature' Ruger's awful factory folding stock. We have them in racks in the cab, but we've never trained to deploy them.

We just had a murder-suicide locally that occurred in the early afternoon.

All three of our beat-cars responded. Our guys exited and found positions of cover as they set up a perimeter. Suddenly, it occurred to one of them that he had a rifle back in the car and that it might be useful. He went back to get it. Neither of the other two officers even remembered the rifles until the situation was over.

The fact that we had never practiced taking the guns in and out of the cars was painfully obvious, and it is therefore no wonder that it never occurred to anybody to actually take the rifle along on the call.

The lesson here is that, at the moment of truth, it is too late to practice!"

Comment: Let us all learn from the foregoing bitter experience, so we don't have to get our people hurt learning what is already known.

15 Sept 2003

Interesting comments from a LEO friend in the Midwest:

"At a recent Simmunitions drill, we exposed officers from the area (including the state capitol) to a situation based on a real incident from a year ago. One of our officers was then interviewing a female domestic battery victim. In the middle of the interview, the suspect (husband) appeared and attacked the victim and our officer with a claw hammer. Our officer dealt with the attack by emptying his canister of OC and then, after it began to take effect, batoning the suspect in the common peroneal nerve on the left leg. A single blow dropped him to the ground. In the meantime, both the victim and the officer were chased around the beat-car by the suspect, who couldn't see well but could charge and still clearly represented a deadly threat!

We reenacted the scenario with a flexible script. Twenty percent of officers hesitated significantly when the victim was attacked via the hammer. Some eventually defaulted to deadly force, but only when the suspect directed his attention at them. It was the collective opinion of the training staff that officers were far too indecisive and far too slow to act when it came to the use of deadly force when its use was clearly indicated.

You might find it interesting that members of the Governor's Security Detail, when presented with the same scenario, all reacted with deadly force immediately. None of them displayed the slightest hesitation, nor vacillation!

Nice to know the governor (at least) is so well protected."

14 Dec 2003

From a friend in WY:

"The Evansville Police Chief had his pistol stolen out of his unoccupied car yesterday while he was in a restaurant in Casper. He told investigators that the pistol was between the driver's seat and the radio console and was 'slightly visible.' It is unclear whether the car was locked or not, but there are no signs of forced entry, no witnesses, and no suspects. The embarrassed chief, of course, 'could not be reached for comment.'"

Comment: If a gun belonging to you falls into unauthorized hands and is subsequently involved in an accident or a crime, and it is determined that your gun had been "inadequately secured" (whatever that means), a legal shit storm will predictably fall right on your head! Leaving "slightly visible" guns in unoccupied cars in restaurant parking lots surely falls into the "inadequately secured" category!

For those of us who carry pistols, it is a much better plan to carry the gun on our person than it is to leave it in cars. The best place for your pistol is on your person where it will actually be of some use to you in an emergency and where you always have direct control over it.

17 Dec 2003

Our naive new mayor in Denver has allowed himself to be pressured by local newspapers on the subject of officer-involved shootings. He is particularly upset when people with knives threaten officers and then get themselves shot. He has decided that nearly all deadly force is unnecessary, and that it can be eliminated with enough application of trendy new equipment and negotiation acumen.

He wants to inundate the department with pepper-ball guns, Tasers, beanbag guns, green strobe lights, OC, sticky stuff, slippery stuff, smelly stuff, stringy stuff, foggy stuff, projectile nets, ad nauseam, all in an effort to eliminate the police use firearms. I wonder how close the mayor wants cops to allow a knife-wielding felon to get before applying an emergency treatment.

Of course, I'm in favor of providing cops with options, but too many options are as bad as too few. Trying even to recall, much less fumble with a bewildering assortment of possible options to which he may have access slows an officer's response and makes him vulnerable in a violent circumstance. It also makes him vulnerable to predictable accusations that he failed to select the best possible alternative. And politicians will be unhappy with anyone who tells them that their pet, new gadget didn't work!

Unfortunately, the public is being told, by naive politicians and journalists alike, that non-lethal/less-lethal alternatives can be used successfully in any confrontation, no matter how violent. Anyone with half a brain knows this is an absurd exaggeration and bandying it about is a disservice to the community.

18 Dec 2003

Firearms Illiteracy in the Press:

In a recent editorial (masquerading as news article) entitled Top Police Gun Prone to Accidental Firing, appearing in the Detroit News, a fearful author, Melvin Claxton, pontificates about the malignant dangers of Glock pistols. It is painfully obvious that Melvin wouldn't be able to distinguish a Glock from a waffle iron if the two were sitting in front of him, but, as is the case with most of the self-important, leftist press, that doesn't hinder him from presuming to tell us all what to do.

With hands wringing, the author laments, "Glock pistols... have earned a reputation among some gun experts as a firearm with too few safety features and that is too quick to fire. Its reputation is directly linked to its design, which ignores important safety features... forces the user to handle the gun with extreme caution."

Well, duh! He fails to name any pistols which can be safely handled carelessly.

He goes on, "The gun's safety features, extremely effective in preventing discharges if the gun is dropped or hit, automatically are turned off every time the trigger is depressed."

Well, duh! I don't know about ya'll, but I want the pistol I carry for security emergencies to actually discharge when the trigger is depressed. What a radical notion!

"'What you have is a gun that is almost too eager to fire,' said Carter Lord (never heard of him), a national firearms and ballistics consultant. 'I think it may be an appropriate weapon for highly trained paramilitary officers in a SWAT team, but not for most police officers and certainly not for civilians.'"

Apparently, no one ever told Mr. Lord or Mr. Claxton that, in America, ALL police are civilians. I'm not sure what country they live in, but in America the police are not a branch of the military. We have civilian police officers here. Additionally, I personally resent the insinuation that all we "most police" are just too stupid to handle a Glock. I wonder what kind of pistol Mr. Lord and Mr. Claxton think we're smart enough to handle!

"Experienced gun handlers, people like former US Border Patrol agent Michael Roth, 66, a small-town sheriff and marksman with extensive gun training" apparently also fell victim to the iniquitous Glock, "In March 1996, Roth was tightening his belt in a mall restroom in Buffalo, NY, when the Glock TUCKED IN HIS WAISTBAND accidentally discharged, striking him in the leg."

Well, duh! With "experts" like this, we surely don't need amateurs.

As if that weren't enough rubbish for one day, we see this headline at the end of the article:

"Weapon easily converted into fully automatic mode" "One of the Glock's most frightening attributes is its ability to easily be converted into a fully automatic weapon capable of firing at the rate of 1,000 rounds a minute."

Of course, Mr. Claxton has never himself even touched, much less fired, any gun in full auto. If he had, he would know how utterly useless a fully automatic pistol is. Indeed, we should pray that all criminals carry full-automatic pistols and use them exclusively in that mode! They would hit nothing and run out of ammunition in less than a second.

Where do these self-righteous cretins come from? The author obviously finds the idea that American police are equipped with modern weapons frightening. He apparently would rather us all be equipped with obsolete weapons.

The willful firearms illiteracy of the leftist press is indeed disturbing, almost as much as their arrogance.

18 Dec 2003

Comments on less-lethal force:

"Since your mayor is such an expert on less-lethal force, maybe he should personally train the entire department, even provide leadership and become the Designated LL Response Officer. Yes, I'd like to see him confronting an aggressive knife wielder with anything less than a firearm. I'm sure he would teach us all a thing or two, about getting killed!

What we have today in police use-of-force situations is rarely 'excessive force,' but rather EXCESSIVELY REPEATED APPLICATIONS OF LESS-THAN-ADEQUATE FORCE by timid and indecisive police officers, allowing a potentially violent situation to spiral out of control."

We cripple our officers with impossible expectations and then naively believe they will be able to protect us. We geld them, and then demand they be fruitful!"

Comment: Chicago PD used to put it this way: "When officers are confronted with unlawful force, they will respond with superior force, until the situation is under control." This seems to have been replaced with, "... officers, fearful of making the 'wrong' choice, will dither around until the situation is out of control." If this is the overall effect of less-lethal equipment, we would be better off with none of it.

We surely need less-lethal options, combined with reasonable expectations. Mayors need to worry less about getting reelected and more about the welfare of their officers and the citizenry.

19 Dec 2003

More information on The Detroit Press' "expert" consultant from a friend who does a lot more expert legal consulting than I do:

"The opposing expert bills himself as a 'gunsmith and firearms consultant.' He was my opposing expert a year ago in a case (involving a Raven pistol in 25Auto, same case as mentioned in a previous Quip) in Florida, which we won and they lost.

The case was a civil suit, brought against the pistol distributor and others (the actual manufacturer, Raven, has been out of business for years). The opposing expert's position was that the diminutive Raven 25, which had been manufactured in the early 1980s, should, by everything that is right and just, have incorporated a built-in locking device. This, despite the fact that there were only two firearms they could name that had ever been made with any kind of built-in lock, both virtually unknown and both long-since out of production, the defunct Fox/Demro 45ACP carbine, and an obscure revolver, at one time made by H&R.

His position was, nonetheless, that a built-in, internal lock could 'easily' have been incorporated in the Raven, and that the pistol was 'inherently defective' for not having such a gizmo. Our position was: 'Oh really! If that's the case, why don't you 'easily' make one yourself, and then bring it here, so we can all examine it?'

He took the challenge, and, over the course of the following year, cobbled together four, successive prototypes of Raven pistols with built-in locks. The sequence of prototypes was necessary, because each, in turn, was shown to have fatal 'problems,' which the expert himself admitted (or, which we pointed out, and he then grudgingly acknowledged). This painful and expensive charade served only to discredit him and his dubious position that the manufacturer could 'easily' have incorporated an effective, built-in lock in the (retail $50.00) pistol. Maybe it wasn't so 'easy,' after all!

At long last, the culmination of this expert's efforts, the final and 'really improved' version of Raven-with-lock, was proudly presented to me at my deposition. I took the contraption, which I had never seen before, and promptly unlocked it using only a paperclip, rather than with the artistically crafted, 'special key,' which was, no doubt, intended to impress a jury (something anyone would be proud to carry on their keyring, to unlock their pistol, in case they suddenly needed to shoot someone) After unlocking the thing with a paperclip, I locked it again and then unlocked it multiple times, using several universally available tools, including a screwdriver. So much for 'high security.' Needless to say, they lost the case!"

Comment: These are the kinds of charlatans who make their living in suits against gun manufacturers, distributors and retailers. We are in desperate need of tort reform in this country!

10. Officer Involved Shootings

29 Jan 2003

From an LEO friend and student on the East Coast:

"One of our officers was just involved in a fatal shooting here. A local convicted felon, out on parole in a stolen automobile, attempted to cause the airbag to deploy on one of our Crown Vic beat cars by suddenly backing into it during a traffic stop. The maneuver didn't work, and the chase was on. The suspect vehicle subsequently suffered tire damage when it ran off the road and slowed to a crawl. The suspect then bailed out, simultaneously firing at our officers (380Auto) over his left shoulder. A single round struck our beat car's open door. No injuries.

Our sergeant, who had just arrived, fired three rounds from his G23 (Gold Dot) at the running suspect. All three rounds found their mark, one through and through on the forearm, and two through the upper portion of his buttocks. The suspect's descending aorta was severed, and he bled-out, posthaste. Neither bullet exited. Suspect was DRT. No one else was hurt."

Lesson: When asked to summarize a recent gunfight, Bill Hickock (known for his short answers) was quoted at saying simply, "He missed. I didn't." Bill correctly concluded that his point was made and that additional details would be superfluous.

There is no substitute for surgical accuracy, no matter how exciting the situation. "Lots of shooting" doesn't end fights.

Hits do!

16 June 2003

From a friend with the LAPD:

"On Thursday, two of our narcotics cops were involved in a shooting in my division. I was the first sergeant on scene.

Two rival gang members had squared-off on a pedestrian-packed sidewalk. Each pulled out a handgun (one a no-name autoloader/mouse gun, the other a Ruger six-inch revolver w/38Spl ammunition) and commenced shooting at each other at a range of ten feet. The Ruger shooter scored six hits! The auto shooter scored only one. Both shooters then calmly walked away from each other.

The Ruger shooter passed his pistol off to a buddy who then turned and faced two of our guys with the revolver still in his hand. Our officers both fired instantly upon seeing the gun. Each officer (armed with Beretta 92Fs) fired two shots. All four shots missed completely! Fortunately for them, the suspect could not have fired anyway, as the revolver was completely expended and had not been reloaded. He dropped to the ground and surrendered without further incident. The shooter himself was arrested a short time later.

The second gang member, the one with the six holes in him, passed his gun to his thirteen-year-old girlfriend, who started to walk away. Two of my sharpest officers pulled up and saw the hand-off. They grabbed her and him and recovered the gun. He went DRT shortly thereafter.

Our Chief Bratton showed up minutes later! (one would never see either of our two former chiefs do anything like that). I briefed him and showed him the scene. The very first thing he wanted to know was that our officers were okay and being taken care of. He made it clear that the suspect's unlucky demise was 'just as well.' I like him already!"

Comment: So do I!

Lessons: Sometimes criminal suspects are competent shooters! We need to be better, faster, and act decisively and without hesitation. At the moment of truth, we will shoot about as well at we did on our worst day of training. The two officers who missed need to take the hint and get to the range!

3 Aug 2003

From an LEO friend in the Midwest:

Officer Involved Shootings

"We had a fatal shooting here on 22 July. The (dead) suspect was probably trying to commit suicide. In any event, he was a construction worker who was hanging around one of our local banks late in the afternoon.

Using a pistol, he kidnaped a bank officer and held him at gunpoint, but then let him go and walked away after only a few minutes when the bank officer explained that he was taking his pregnant wife to the hospital (The story was true! The bank officer's wife delivered a healthy baby boy later that evening)

Four of our officers soon located the suspect as he walked away from the bank building. As they commanded him to stop, he turned to face them with the pistol in his hand, and all four officers simultaneously fired their pistols at him. Range was eight meters. A total of nineteen shots were fired at the suspect by our four officers. Of nineteen, only four hit, and all four were fired by the same officer. All fifteen shots fired by the other three officers missed. Suspect was struck in the torso. He took a few steps backward and collapsed. He went DRT within a minute.

The officer who did all the hitting is one of our range officers and an extremely competent shooter. Not known for his speed, he is known for his deadly precision. He was shooting a G21. I don't know the brand of ammunition. All other officers were shooting G19s.

Subsequent investigation revealed that one of the participating officers (who missed) had not been to our shooting range in over a year. We're all now trying to figure out how that could have happened!

All officers have been cleared by the DA, and no one (aside from the suspect) was hurt, although several of our errant bullets did cause minor property damage. Most were not found. The suspect never fired a shot."

Comment: Of four ostensibly trained police officers, only one was competent enough to decisively end this fight. He carried the fight for the other three and may have saved their lives. To be sure, all were courageous for answering the call and for being there, but courage does not substitute for competence. Personal competence cannot be acquired without personal effort and personal commitment. The department can provide the trappings, but the individual must provide the requiem personal devotion and determination. On this group of four, only one did.

18 Oct 2002

Shooting incident in SA:

"One of our officers was involved in a shooting today. He was responding to a call for assistance. On arrival, he was confronted by an ax-wielding EDP who was attacking passing motor vehicles. Our officer verbally challenged the EDP. The EDP jumped off a truck he had struck several times and advanced towards our officer.

Our officer had left himself sufficient time and space (per his training). Our officer, realizing that he would probably have to shoot, lined himself up so that the truck was direct in his line of fire and would serve as a backstop.

Sure enough, the EDP put his head down and charged with his ax raised. Our officer fired two rounds (9mm 115g FMJ) from his CZ 75. Both hit in the upper part of the EDP's legs, causing him to pitch forward and fall to the ground. Our officer then stepped in and kicked the ax away.

I interviewed the officer two hours after the incident. He is one of my (and your) students. He reported that, at no time was he unsettled. He stated that all of the elements we teach, scanning, moving laterally, verbally challenging, etc came into play. He said it "all came together" and worked well (big smile when he said that).

Asked why he hit the EDP in the legs, he stated that the EDP was running toward him with his head down. So, he started his 'zipper technique," but stopped firing as soon as his target went down and presented no additional threat.

He stated that he can clearly remember using his sights. 'That's the way I was trained,' he said, matter-of-factly."

Officer Involved Shootings

Lesson: Let your opponent panic. When you move, verbalize, and shoot accurately (using your sights), you are pretty hard to beat (even when you're only shooting 9mm hardball).

11 Oct 2003

One of my LEO students in a large city in the Midwest was involved in a fatal shooting last week. He is an exceedingly competent shooter and a dedicated trainer. His skills were tested:

"My young partner (two weeks out of the academy) and I responded to a domestic call. Our department just graduated a large academy class, and all us senior patrolmen are currently functioning as FTOs. A man had accused his wife/live-in of hiding his drug supply (crack cocaine) from him. He subsequently became angry and threatened her with a pistol (Bryco 9mm, bright chrome plated). The call to 911 was made by the woman. The offender, upon discovering that his wife had called the police, said that he would 'have a little surprise for them (responding police officers) when they arrived.' He then went outside and waited for us.

He didn't have to wait long! The two of us arrived, parked one house away, and approached the house in question on foot. We saw the suspect standing near the sidewalk, but it was a warm evening, and there were many other people walking about, as well as a good deal of traffic. However, the offender's stance (his hands were not visible) made me particularly suspicions of him. I said to my partner, 'See the way that guy is standing? That may be our suspect.'

He waited for us to get within twenty feet. We were commanding him to move slowly and show us his hands. Without saying a word, he brought up the Bryco pistol and pointed it directly me. I responded by lurching to the side and simultaneously drawing my SIG P220 (my hand was already on it). As soon as I found my front sight on his body midline, I fired several rounds (230gr Gold Dot). I could see his shirt convulsing, so I knew I was hitting him.

He stumbled backward but stayed on his feet and did not drop the gun. I could see him frantically yanking on the trigger of his shiny pistol. He was trying desperately to shoot me. I fired several more rounds, again into the body midline. By this time, my partner was also firing (9mm, 147gr WW). The suspect fell backward onto his fanny and then fell the rest of the way, hitting the back of his head on the ground. However, he still had not dropped his pistol.

Then, he sat back up and pointed his pistol at us once more! I knew I couldn't have more than one or two rounds left in my magazine, so I put my front sight on his head and fired what turned out to be my last two rounds (shooting him in the pelvis would have been pointless, as he was already sitting). One round hit him in the throat, and the other hit him in the eye. That did it! He finally dropped the pistol and fell back down, DRT.

I reloaded but have no memory of it. My partner commented on my fast reload. When he made his comments, I didn't even realize that I had reloaded.

Of nine shots I fired, seven struck the suspect. Of the seven that hit, two, fully expanded, still went through and through. The rest stayed in the body.

As it turns out, the suspect never fired a round. His pistol had a fully charged magazine inserted, but there was no round in the chamber. Either through ignorance or carelessness, he had not loaded his pistol. He brought a club to a gunfight. Shame on him. He won't have the opportunity to make that mistake again!

The department backed us fully. What little criticism there was of our actions was handled deftly, professionally, and quickly."

Lessons:

> There is no "safe way" to walk up to a suspect, just as there is no "safe way" to approach a vehicle. In situations like this, there is seldom going to be useable cover available, so sudden, lateral movement, combined with

a smooth draw, and decisive, aggressive shooting will be the difference between you living or dying.

> In gunfights, you're going to miss with some of your shots. No one bats a thousand in this business. The important thing is to finish the fight! We carry pistols because they are convenient, not because they are effective. Multiple shots will probably be required, maybe even an emergency reload, or two. All this must be practiced thoroughly. If there must be a fight, the best one is a short one. Deadly force should always be applied with surgical precision but also with great enthusiasm, sufficient volume, and without hesitation nor apology.

> Even when a suspect is off his feet and on the ground, he can still be dangerous. You have to watch his hands. When he still has a weapon, make your judgment based upon his capabilities, not his "intent," nor on his posture.

> My students were victorious because of skill, teamwork, preparedness, good equipment, personal gallantry, and superiority of purpose. Good show, guys!

21 Oct 2003

A success story from one of my students who is the training officer with a Midwest police department:

"Last night, one of my officers responded to a violent domestic, reported by a neighbor. When my officer arrived, he heard loud, female screams from inside the house. When he entered, he was immediately confronted by a large man who had a knife in his hand. The suspect slashed his own arm, then turned to my officer.

My officer, pistol (G22, with 165gr Gold Dot) in hand, ordered him to drop the knife several times. The suspect not only refused but moved aggressively toward my officer. My officer responded by retreating and repeating his command to drop the knife. The suspect was quickly overtaking my officer when my officer decided to fire. He fired a single shot, which struck the suspect in the center of his abdomen.

The effect was dramatic and immediate! The suspect fell backwards, landing on his backside. In the process, he dropped the knife and pleaded with the officer not to shoot him any more. Backup arrived shortly, and the suspect was arrested and transported to the local hospital. He is expected to survive. My officer was unhurt, and no one else in the house was hurt.

The single round fired by my officer penetrated the suspect's abdomen, ricocheted off of the pelvis, and then exited out his side. It then (fully expanded) reentered his right arm and finally came to rest just under the skin on the opposite side of the arm. It performed as advertised! I quickly went to the location myself and personally took the officer to the PD in order to get him away from the scene. The involved officer was, of course, suspended (with pay) pending the completion of the investigation, but none of us see any problem with it.

The important thing was that we have an enlightened policy here, so that this officer was not forced to make any statement in the immediate aftermath of the incident. He was taken home with his pistol and his patrol car and was not demeaned by having these two items taken from him in public.

I now understand how you feel when one of the students you have taught is victorious in a deadly encounter. As one of your students, I have passed on what I have learned to my guys. I can't tell you how good it feels to know that all my (and your) hard work has paid off last night."

Lesson: No deadly-force episode, no matter what the circumstances, will ever be exactly duplicated, before or since. And, no matter how restrained and righteous the officer's actions, someone will always point out where he could have done it better. Happily, the law doesn't require any of us to be saints. The law requires us to be reasonable.

An enlightened department will thoroughly, and with no bias, investigate all deadly-force incidents involving its officers. However, enlightened departments also don't reflexively treat their officers like criminals every time those officers are obligated to make difficult decisions quickly.

Good show, both of you!

Part 3: Weapons & Equipment

11. Ammo

19 Jan 2003

A student and good friend in a course we are doing here in CA, like me, uses a G32 in 357SIG. His slide cracked at 10,000 rounds and was promptly replaced by Glock, at no cost to him. Yesterday, at 20,000 rounds, the barrel lugs sheared. I'm sure Glock will take care of that too, but this caliber (357Sig) is obviously a lot harder on guns than is the 9mm.

High chamber pressure and high slide velocities obviously take their toll. My G32 is at 2,500 rounds (mostly Cor-Bon), and I have had no problem with it. I also have a SIG239 in 357Sig. It is new and works fine. We'll see how it holds up as we use it.

I believe the 357Sig caliber to be a significant development in handgun technology. From what we know so far, its terminal effect is outstanding, but guns chambered for it, even Glocks, are going to be battered as they are with no other caliber. Glock's and SIG's excellent service philosophy are surely helpful, but the user's expectations with regard to the useful life of the pistol are going to have to be adjusted accordingly.

23 Feb 2003

From an LEO friend in WI:

"We recently acquired some Mexican Aguila 'IQ' 9mm rounds. They are a 65 gr HP. We heard that they would penetrate soft body armor, so we tested them. We shot them into fifteen-year-old, retired Second Chance vests. We used a Taurus pistol with a five-inch barrel.

Range was three meters.

No penetration. Not even close. Only the first two layers were penetrated, and deformation was minimal. This old vest also stopped a dozen other brands of 9mm rounds we shot at it, including all American manufacturers."

Guns & Warriors: DTI Quips – Volume 2

Lesson: Don't get excited about any of the "magic bullet" rumors that periodically circulate. No one has repealed any laws of physics recently.

1 Apr 2003

A friend overseas on the subject of shotgun flechette rounds:

"Twelve-gauge 'flechette' rounds are not recommended. Compared with 00 buckshot, their fatality potential, at least at close range, is similar, but stopping effect is disappointing. Buckshot takes people down much faster and is effective at a greater range. Flechette rounds have come and gone here!"

Comment: Like most fads, shotgun 'flechette' ammunition is inferior to what we're already using. Don't waste your money.

19 May 03

Of Reloads and Glocks…

At a course in Michigan last weekend, we had a case rupture in a Glock 22. Gas pressure blew out the slide-lock lever and half the trigger. The shooter received a gas cut on his hand but was not otherwise seriously injured. With the addition of some Band-Aids, he continued with the course. With the replacement of the trigger and slide-lock lever, the pistol can probably be returned to service.

In my judgment, the case in question had been reloaded one time too many. It blew out at the unsupported portion, just as one would expect. I believe that progressive brass flow toward the front had thinned out the rear of the case, to the point that it could no longer contain the pressure.

Reloading 40S&W cases, even ones that have only been fired once, may be a bad idea. Nine millimeter and 45ACP cases can be reloaded, it seems, numerous times with little concern. Not so with the 40S&W, and particularly not so with the 357SIG. I don't recommend reloading either. Better that fired cases be discarded (and you thus shoot only new, factory ammunition) than you see you pistol ruined.

23 May 03

More about the performance of the M16/62gr "penetrator" round:

"After-action interviews with infantry Marines indicate that most small-arms engagements occurred within 100m, many within 25m, even in this desert environment. Jihadists tried to abrogate our artillery, helicopters, and CAS by hiding until we were nearly on top of them and then attempting to ambush our lead elements. However, their individual marksmanship was poor and their movements sluggish. Our response was aggressive and rapid. They were, in every case, summarily outflanked and destroyed.

Inadequacy of our 5.56 ammunition is well known among the troops. This is evidenced by the large number of head shots, at all ranges. Our Marines can shoot and hit, but many decided not to take chances and opted for a head shot.

That is, of course, well and good, but many of us are still pushing for a rifle round with improved penetration, range, and terminal effect. As you said, we want it all in place before the next war!"

2025 Update: Twenty-two years later, the successor to the 5.56x45 (223) is a necked-down version of the 308, in a hyper-pressure, bi-metal case. Progress has been slow, and this new round is still not been tested in actual fighting. There are many issues, and the jury is still out!

12. Pistol Advice

4 June 2003

At a LEO Program in Michigan earlier this week, we had a small-statured student who was using a department G22. He had before never shot so many rounds in such a short time.

After several hundred rounds of normal cycling, his pistol started to suffer from failure to eject. The problem grew progressively worse, until I suggested to him that he use another gun. However, all the G22s he tried soon yielded the same result.

He was stocky but had small hands and could not grasp the Glock so that the bones of his arm were directly behind the frame. That was the problem!

I suggested he switch to a G23 with a ROBAR Grip Reduction. I had one with me that he was invited to use. The increased slide velocity, combined with the smaller grip solved the problem completely, and he was able to finish the course with no further gun/functional problems.

The lesson here is that equipment has to fit the individual. People with small hands invariably have trouble with double-column pistols, particularly those with light, polymer frames. They may need a single-column pistol or a grip reduction. Happily, we were able to convince the department training officer, who was there also, that the existing department gun (G22) was unsuitable for this particular student.

Happy ending!

2025 Update: Reputable pistol manufacturers now offer variable grip-geometry with most of their pistols, making the task of acquiring a correct grip much easier for many

6 June 2003

At an LEO course in MI earlier this week, we scheduled an extensive low-light shooting session with handguns. One department's representatives had their G22s all equipped with M3 quick detachable flashlights.

One student observed:

"The light is always in alignment with the barrel and reloads can be done without tucking the flashlight under one's arm. Shooters can keep both hands on their pistol when shooting (unlike the case when using the Harries' technique) and, even with one hand injured or occupied, the shooter can still use gun and light together.

However, the light is always in direct alignment with the barrel, so one can't light up something without pointing the gun at it. Also, because the light is right below the barrel, when one takes more than four shots from any one place, he will discover himself in a 'white out,' where the light is hitting smoke and unburned powder hanging in the air, effectively masking the target. No problem if there is room to move, but an issue in tight quarters.

The user must learn how to attach and remove the flashlight while keeping his hand below the line of the barrel and must also learn how to avoid the 'constant on' switch."

Comment: I've already editorialized on the subject of multipurpose emergency equipment. Personally, I prefer a gun that is just a gun.

7 Aug 2003

Yesterday, I handled a S&W 1911 at a retail gun shop in Nebraska. Like the one I handled in Orlando, and unlike the one my friend handled in Colorado, this one was slick, and everything on it worked well. I'm making it a point to handle everyone I can. S&W is obviously working on quality control and deserves credit for that. The one in Colorado may have been an aberration.

This same gun shop had for sale several SigPro pistols with a two-position, slide-mounted, manual safety! I had never seen this arrangement before.

The pistols also had the traditional, SIG single-stage decocking lever. The clerk explained to me that SIG made a run of pistols for a foreign government, and the buyer insisted that SIG install a manual safety lever, even though a manual safety on this particular pistol is a silly redundancy.

The safety lever faces forward and must be pushed down to get it into the "off" position. Only the largest of hands can accomplish this maneuver without compromising the master grip, so, for most shooters, this manual safety isn't useable as such. This is an example of bureaucrats, who know nothing about guns or fighting (and couldn't care less), buying serious, defensive firearms for police officers who will actually have to use them.

We've had many SigPros in courses, and, like all SIG pistols, they work just fine. These special ones were all reduced in price by $100.00. If I owned one, I would simply leave the manual safety in the "off" position. I sincerely hope that is what the ultimate users of these pistols do.

10 Sept 2003

SIG's "K" Trigger Update

Last weekend, I got the chance to run live rounds through a SIG 229 (40S&W) that was equipped with a "K" trigger. I expressed the opinion that this new system will quickly displace the manual decocking lever that most SIGs come with now, and my opinion was not changed. SIG is now taking police department orders. The K-trigger will be available to the general market in 2004.

The K-trigger will be offered as an option on the 226 and 229 (which will be called the 226DAK and the 229DAK). Single-column pistols, like the 239 likely will be eligible for the new trigger inside of a year, as modifications need to be done to the inside of the frame. The new trigger will not be available as a retrofit, as, again, the frame needs to be modified. It will be offered on new guns only.

The K-Trigger is short and smooth at a consistent 6.6lbs. It is not a two-stage trigger like the one founds in Glocks. There are two links (trigger reset points). The shallow one yields a slightly heavier press (7.5lbs). The deep link re-presses at the same 6.6lbs. The point is that the pistol will fire from either link, and it will drop the hammer for a second time on a dud round (from the second link). Hammer spur is gone (along with the decocking lever), and the system can be used in conjunction with either a regular or a short trigger.

The traditional SIG self-decocking system features a long trigger pull at a constant, eleven pounds. It is called (appropriately) a "flat revolver." That system will continue to be produced, although the K-trigger, as opined above, will quickly supersede it. In fact, the K-trigger will also supersede SIG's manually decocking system.

I will have and carry a 229 in 357SIG with the K-trigger shortly and will get a chance to give it a thorough workout over the next number of months. I am convinced now that, when SIG markets this new system aggressively, they will significantly dig into Glock's market share.

2025 Update:

SIG pistols mentioned above have since been mostly superseded by the striker-fired SIG 320 and the smaller (also striker-fired) SIG 365. In 2023, the 320 (M11) was selected by DOD as the successor to the Beretta 92FS (M9). However, the future of the 320 is, as of this writing, in doubt! Mysterious "uncommanded discharges" have plagued 320's owners, to the point where police departments, even military units, are currently dropping it from issue/use.

28 Oct 2003

The Critical Combination:

At a pistol course in Texas last weekend, I had a female shooter with a G19 who was experiencing one stoppage after another. Most involved the slide failing to go completely into battery. Lubrication helped, but the problem returned within an hour.

The woman was small statured, even for a female, and she indicated that her pistol normally works flawlessly and that she could not understand why it was not working well this particular day. When she shot a G32 and a G23, no functional issues manifested themselves. Both worked perfectly in her hands. It was only her G19 that experienced the problem.

I explained to here that she had encountered the "Critical Combination:"

(1) a G19,
(2) in the hands of a small-statured person,
(3) shooting wimpy ammunition.

The 9mm hardball ammunition she was using at our course was foreign-made and wimpy. When she shot high-performance, service ammunition she had, of course, not encountered the problem.

The G19 is a fine carry gun, but it needs to be fed full-power ammunition. It will usually still work, even with wimpy ammunition, in the hands of an average-to-large-sized male. But small-statured shooters will invariably experience problems when they attempt to shoot weak ammunition through it.

3 Nov 2003

I got my hands on a G37 at the new Cabellas store in Kansas City today. It is slightly fatter than a G17/22/31, but not nearly as big as a G21. If it catches on, I'm sure we'll see G38, which will be similar in size to a G19/23/32. Magazine is staggered and holds ten rounds. I suspect the LEO version won't hold any more.

Caliber is listed as "45GAP" (Glock Automatic Pistol). Curiously, Cabellas didn't have any ammunition for it, nor were they sure when they would get any in! They didn't even know who (if anyone) was making it.

The 45GAP round is slightly shorter than the 45ACP, so a 45ACP round would not allow the slide of a G37 to close if one mistakenly finds its way into a G37 magazine. The 45GAP probably will chamber and fire in most pistols chambered for 45ACP, but I doubt that it would hurt anything. Much like firing a 38SPl in a 357Mg revolver, although, in the case of the 45ACP, the 45GAP may not cycle the slide.

So far, the G37 has not garnered much popularity. Like all Glocks, I'm confident it works just fine, but I'm not sure what it is for or what market niche it is supposed to fill. In any event, this is the first one I've seen in a gun shop counter. The clerk indicated interest was scant. In fact, I was the first one in a week who had asked to see it. Not surprising when ammunition is unavailable!

2025 Update. The 45GAP pistol caliber has since died! Not even Glock could keep it alive. Unavailable today.

4 Nov 2003

Comments on the G37 from large retailers:

"I've actually had 45 GAP ammo (Speer) for longer than I've had the pistols to fire it. I've sold a few boxes of the ammo to curiosity seekers but have yet to find anyone who wants the pistol. I'm baffled by this pistol and cartridge.

(1) It produces ballistics nearly identical to the 40S&W (Glock 22)

(2) It is enough thicker that no currently available holster will accept it, and

(3) It 'features' lower capacity magazines than the G22, at least in the LEO version.

My honest suspicion is that Glock wanted a cartridge that bears the Glock name, so a problem (and corresponding 'solution') were suddenly 'invented.' Can't have a 40S&W, a 357SIG, and a 400Cor-Bon and not have a _____ Glock, eh?

Other dealers I've talked with have had identical experiences."

"Speer is actually making two loads, a cheap FMJ and a pricey Gold Dot.

Winchester just started cranking it out today, supposedly."

27 Dec 2003

I visited a large, local retailer today. I asked him about Kel-Tec pistols, specifically their compact 32 and 380 autos. He indicated that he was unable to keep the 380 in stock and that both had shown themselves to be reliable and easy to use. He went on to say that these small Kel-Tec pistols sell better and have far fewer problems than the Beretta Tomcat, the Taurus equivalent, or the NAA equivalent, combined. In addition, the Kel-Tec is thinner and lighter than any of them.

I've seen several of these small pistols in courses, and all have worked just fine, but I have been doubtful with regard to their actual usefulness. However, I have to admit that everyone who owns one thinks they're great. One can surely carry one in many places where most other pistols would not fit. I may start carrying the 380 auto version as a backup!

28 Dec 2003

Comments on Kel-Tec pistols:

"The pistol is reliable, far superior in that regard to the PPK or TPH. The slide is difficult to cycle, and I wish it had better sights, but understand why things are the way they are."

"The trigger reset on these little guns requires travel back past two 'clicks.' If you only release the trigger past the first 'click' you will cause a failure-to-fire malfunction, requiring a tap-rack-bang drill to remedy."

"The ejector throws cases directly at your face."

"Kel-Tec is an innovative company, and their customer service is second to none."

"I have carried my Kel-Tec P32 for over three years as a backup pistol. It is so small and light as to be unnoticeable."

"A problem with the trigger bar caused me to return my pistol to the factory, where it was totally rebuilt: new slide, barrel and internals, and a new magazine. It has not malfunctioned since. NO COST, and they even sent me a check to cover the cost of sending the pistol FEDEX overnight. Kel-Tec's customer service is top drawer!"

"They rarely last more than a few hundred rounds without needing repairs. As with most guns, a typical retail purchaser never shoots more than one magazine through it during the entire time he owns the gun anyway."

"I've never had a single purchaser unhappy with a Kel-Tec pistol. They have all functioned flawlessly."

"My Kahr P9 Covert in a Ky-Tac Pockit Lockit will fit in the same space as a Kel-Tec 380, but offers 9mm Parabelum power, the 'minimum acceptable' level."

"Kel-Tecs are known here as 'Jam-Tecs.' They are ammunition sensitive and are notorious for breaking triggers and extractors. I realize some people love them. I don't."

"At 750 rds, the trigger stopped actuating the striker. I returned the pistol to the company. They repaired and returned it to me, all in four days, and at no cost. I now have an additional 1000 rds through the gun without any failures."

My comments: Mixed reviews, but there is a consensus on Kel-Tec's Customer service. It is good. After I've had a chance to carry and use mine for a while, I'll have more.

13. Pistol Reliability

18 Oct 2002

We just completed a Basic Pistol Course in PA. Most of the students were supplied with G19s and G17s. One brought his own Beretta 92F. Another brought a Walther PPK/S. The Beretta and its owner held up fine. Glocks, of course, all did well. The PPK/S lasted for about twenty minutes. Too many feeding and other functional problems (hardball). The student gave it up and went to a Glock, which worked fine for the rest of the course.

Lesson: PPKs may work for James Bond, but I've seen precious few copies I'd give a dime for.

25 Feb 2003

A Glock Story from a friend in WA who, like me, carries a G32 (357SIG):

"Per your suggestion, I called Glock shortly after our class. They asked for the gun to be sent to them and for a history. I sent them the gun and all the details of the broken recoil spring and other parts. Glock's reply was frank. They indicated that I was way out on the end of the bell curve for the predicted useful life of the pistol, 20,000 rounds. They asked to keep it the pistol for a detailed engineering evaluation.

This is what they sent me in return (at no cost to me):

>A brand new G32.

>A 40S&W barrel that essentially transforms the gun (at my option) into a G23.

>Free admission to the Glock Armorer's Course

>Miscellaneous spare parts and promotional items.

As always, Glock took care of it.

Comment: The 357SIG is indeed an exciting development in defensive pistol ballistics. I really like the cartridge and its ballistics.

But, there is a price to be paid. Slide velocities are such that the life of the gun is going to be greatly reduced from that of the same gun in 9mm. The useful life of a G19 is essentially unlimited, but 50,000 rounds at a minimum. The same gun in 357SIG (G32) is probably limited to 5,000 rounds. It's just the ballistic facts of life.

Glock's customer service is, as always, wonderful. They take excellent care of their customers, and they deserve a lot of credit for that.

30 Apr 2003

From an LEO friend:

"Had to fire on a pit bull a couple weeks ago. My Taurus model 85 (38Spl) failed to fire the first four times (striker pin struck off center and not hard enough to detonate primer). On my fifth pull of trigger, it finally fired and scared off dog (am ashamed to say I missed).

I was scared and, needless to say, unhappy with the gun. The next day I sent it back to Taurus with a letter. I got the repaired gun back, with no response. When I called, the person I talked with explained that they had replaced three parts, cylinder locking pin and spring as well as the firing pin spring. He went on to say that my pistol had failed "due to excessive use".

This is a titanium revolver with fewer than 150 rounds through it. I had been carrying it in an ankle holster, but no more. I traded it in the next day on a scandium S&W 340PD.

Good thing that dog didn't charge!"

Comment: Taurus is surely not going to make many loyal customers this way!

2025 Update: Taurus has come a long way since 2003. Today, their quality is much improved.

Pistol Reliability

7 May 03

"Normal Capacity" magazines not available for the XD Pistol:

A friend in the magazine business just informed me that normal-capacity Beretta magazines have been modified to fit in the XD pistol, but that they don't work well and are not recommended. Since the XD came onto the market after 1994, no non-restricted "normal capacity" magazines have ever been available from the factory. All magazines are reduced capacity.

As I mentioned before, this is not a big issue with the compact models, but it is with the full-sized ones. Normal capacity, factory Glock magazines, in both 40S&W and 9mm, are still commonly available in the aftermarket, and I don't think the XD will ever be able to compete effectively for the business of serious students of defensive pistolcraft when all they can offer them is an eleven-shooter. Glock users, in the interim, will carry thirteen (40S&W) and seventeen (9mm) shooters.

This may all change next year when the 1994 Crime Bill sunsets, assuming our president is able to find his backbone and allow it to do so.

2025 Update: It did!

10 May 03

From a student:

"I am entering the Cook County (IL) Sheriff's Police Academy. I planned on buying and using a G23 (40S&W). I have been informed that Glocks and H&K USPs are not allowed, only S&W, SIG, Beretta, and Ruger. Only calibers allowed are 9mm and 45ACP; no 40S&W and no 357SIG."

Comment: Some police officials bring their personal political agendas to the job and see nothing wrong with imposing them on everyone else. Someone entrenched at the Cook County SO "just doesn't like Glocks," and it will be that way until he leaves or is (finally) overruled.

"Men occasionally stumble over the truth, but most of them pick themselves

up and hurry off as if nothing happened."

Winston Churchill

22 May 03

Last night I had a conversation with the head trainer of a large state agency in the Midwest. Our discussion was about service handguns. When I first started working with this agency a number of years ago, their agents (all plain clothes) carried S&W 4516s. The act of manual decocking was, of course, included in the training curriculum, and was accepted by all. In recent years, the agency, once independent, has come under the patronization of the State Police.

State police officers here, nearly all uniformed, carry SIG 229s (40 S&W), in SIG's current self-decocking (DAO) configuration. The head of the State Police insisted that this smaller agency get rid of all other handguns and also adopt the SIG pistol. They duly obeyed the edict, and they all now have SIG pistols in 40S&W. My friend was able to successfully lobby for the smaller M239, but the self-decocking configuration was declared "not negotiable."

S&W, Beretta, and SIG all offer self-decocking versions of their autoloading pistols. Glock was, of course, the first self-decocking pistol, and, with no manual safety levers, nor decocking levers, and no safety/decocking levers, its popularity in the law enforcement business speaks for itself. The Glock trigger features the best of both worlds: it provides the shooter with a shallow reset for fast and accurate follow-up shots, but it still decocks itself as soon as the trigger is released completely. Manual decocking is eliminated. S&W's, Berettas, and SIG's self-decocking versions have, until recently, provided only a long, heavy pull and a subsequent long reset for all shots. In fact, agents superciliously refer to SIG's version as merely a "flat revolver," which is actually an accurate description.

Manual decocking and manual safety levers have become unpopular among police executives. I do not see that trend reversing any time soon.

9 June 2003

Pistol performance in the Philippines, from a friend there:

"Over here, Beretta 92s have endured only because those who have them don't shoot them much, far less than you Americans. Some have ditched them because of grip size. We have small hands compared with yours.

CZ75s and clones fare slightly better. Rarely do operators carry them hammer down on a loaded chamber. Cocked and locked is the preferred mode, due (once again) to trigger reach. In terms of durability, frame, slide and barrel of the CZ's have stood up to hard use, but smaller, internal parts have not. You need two to keep one running.

SIGs have not seen real use here, again due to trigger-reach difficulties. They're just not designed for our small hands.

Kahrs have a small but fiercely loyal following.

Glocks have suffered a bad rap mainly due to slide breakages, either near the ejection port or under the muzzle. It doesn't happen often, but it happens often enough to be alarming to some. I'm not sure why it happens here and not in the USA.

Taurus PT92s (and all other pistols with three-position safety/decocking levers) have developed a bad rap here as well. Your African friend's observation is common here. Many users find that, when the pucker factor accelerates, they bear down on the safety/decocking lever so hard that the gun cannot be made to fire.

The Walther P99 developed a strong patronage, but, seven years later, we're having problems with broken firing pins, even when snap caps are used to dry fire. In addition, sights routinely wobble in their dovetails.

The 1911 lives on, but it too isn't free from woes. Firing pin stops work loose. Extractors require frequent checks for proper tension (and breakage). Slide stops and hammers break, and improperly adjusted internals often turn the gun into a machine pistol.

My own two cents reflects the advice you give in your rifle book. You may not find one gun that has all the ideal features, but you should come close if you shop hard enough. In our case, it's a little trickier since our hand sizes are significantly smaller than those of our American or European brethren."

Comment: A piece of emergency/safety equipment as personal as a defensive handgun needs to fit properly and suit the user. No one gun will "fit all," despite the efforts of big police departments to pretend it is so. The search will last a lifetime (which we all hope is long and prosperous!)

14. Pistol Problems

7 June 2003

On the Taurus PT92 safety/decocking lever, from a friend in Africa:

"We had a Taurus PT92 in a course, a shiny stainless number! This gun is basically a Beretta 92, with a frame-mounted 'manual (three-position) safety/decocker'.

When carried cocked and locked, pushing the safety into the 'off' position too enthusiastically also decocks the pistol, negating the benefit of having it cocked in the first place. In addition, when the shooter then holds the lever down, the pistol won't shoot at all!

The student using this gun was an 'expert' in double-action/confusion shooting by yesterday evening.

No thanks! Give me a Glock any day!"

Lesson: At the moment of truth, the last thing you need is a bewildering choice of several ways to proceed. This is why pistols like the PT92 have never gained any real market share and why pistols like Glock seem to have cornered it all. The Glock has only one component to its fire control system, the trigger. No manual decocking levers; no manual safeties. Nothing needs to be done to it in order to enable it to shoot, and nothing needs to be done to it after it is shot in order to get it into a condition where it is appropriate to holster. The gun does everything for you, except shoot itself. No choices. No confusion.

19 July 2003

A NY trigger saves a life, from an LEO friend and student in SA:

"I've resisted the idea of the NY Trigger, but, after a recent ND with my G26 (a finger inadvertently entered the trigger guard as I was picking it up) I installed one on the same G26. That was several weeks ago. I carry this gun as a back-up to my issue Z88 (Beretta 92 clone). Off duty, I carry the G26 only.

Sure enough, earlier this week I confronted a kapmes-wielding drunk (a 'kapmes' is a long handled machete, common here) who was threatening me and several others standing in the area. I drew my G26 and, after picking up the front sight, I warned to suspect verbally several times.

No response.

I started to press off the first shot, but, just as I hit the link, I saw the suspect dropping his weapon and putting his hands up. I immediately got my trigger finger back into register. When help arrived, we arrested the drunk.

That NY Trigger definitely saved his life. No doubt!"

Comment: Of course I don't know all the details, but the foregoing is one of the reasons I recommend the NY Trigger to most of my students who carry Glocks. I have a NY Trigger installed on all my Glocks.

I've surely seen pistol triggers that were too heavy for practical, defensive purposes, but an eight-pound break with a six pound take-up is not too much of an encumbrance for most shooters.

2025 Update: Current generations of Glock pistols have corrected all the issues that made the NY Trigger something I recommended.

NY Triggers are no longer necessary, nor relevant.

22 July 2003

Poor Marks for S&W from a Friend and Student:

"After your glowing report on the new S&W 1911 that you shot at the recent IALEFI Convention in Orlando, FL, I went yesterday to a big gun store with which we are both familiar. I had every intention of purchasing one. They had a copy under the counter. The clerk, who is a good friend, just rolled his eyes and handed it to me.

What a disappointment!

The plastic butt pad on the Wilson magazine (supplied with the pistol) promptly fell off as soon as I picked it up. The feed ramp is the roughest I have ever seen on a 1911, I mean serious, easily visible, uneven, ridges. The trigger is heavy and gritty.

Worst of all, the manual safety lever has a clear and distinct 'false on' position, short of the true 'on' position. This isn't just a hesitation or a rough spot. It is a definite 'false stop.' Moreover, it took all the strength I could muster with both my thumbs to force the safety lever up to the true 'on' position.

A good gunsmith could, I'm sure, easily fix all the problems I saw, but how did this gun ever get out the door? I had a boyish hope we might finally have a gun from this manufacturer that was ready for duty, right out of the box.

With all due respect, my friend, you need to look at more than one before you write glowing reports!"

Comment: My friend is right, of course. I obviously spoke too soon. Poor quality control has been S&W's nemesis for decades, and it appears to have reared its ugly head once more, just as the company was emerging from its self-imposed public relations disaster.

It is so frustrating. We all want this grand old American company to succeed! But, with issues as described in the foregoing, success will continue to elude them.

18 Nov 2003.

On Glock maintenance from an armorer in a large PD:

"One of our deputies was qualifying last week (G21). He had four failures to fire within a single magazine. He, of course, performed a T-R-B drill each time, and managed to get through the magazine, but he ejected four live rounds onto the ground in the process. All four had dented primers, but the firing pin hits appeared light.

He was directed to one of the department armorers (me). I discovered so much 'gunk' inside the firing pin hole, that the motion of the firing pin was retarded sufficiently to cause misfires. Our department prohibits our deputies from disassembling their Glocks beyond the frame, slide, barrel, and recoil spring. This deputy said that he thought 'someone' had recommended that he 'lubricate' the firing pin often. He had performed this 'maintenance procedure' religiously.

We cleaned the G21 up, and it ran fine. However, I am concerned. This deputy should have never been told to oil the firing pin, and there needs to be a system here through which duty weapons are completely broken down at least once per year in order to scrutinize and clean those areas that are rarely inspected.

I've offered to perform complete weapons maintenance for my station, but I only work part time, and the department doesn't want to spend the money."

Comment: Never depend on anyone else to "automatically" maintain your weapons. You must take personal responsibility for your own safety. In every sense of the word, you're on your own!

15. Rifle Advice

18 Oct 2002

Advice from a friend in the federal system, from a recent lecture:

"If you're not familiar with the AR-15/M4 rifle, you should be. I recommend that you become familiar with how to load and fire it. Also, become familiar with the AK-47. Sooner or later, you'll run into these weapons, and you need to know how to use them.

When it goes bad, it goes bad fast, very fast! You'll have only a short window through which to take dynamic, positive action. Don't hesitate and don't miss!

You'll have to move, move fast, and keep moving. A moving target is a difficult target.

Let this be your law: Prepared for anything, depending on nothing.

Think in terms of being self-contained. Don't deceive yourself into thinking that someone is going to rescue you in the nick of time. That only happens in the movies!"

28 Nov 2002

More on military rifles from another international friend:

My brother has just arrived in Kiev, Russia. He reports that the new Russian AN94 is a good and functional rifle. Most African countries are now buying Russian small arms. Price, durability, and ease of use are big draws. At less than one hundred dollars per copy (including four magazines) no western rifle comes close to the AN94. Russian rifles are rude, crude, and not particularly accurate (at least by Western standards), but they are designed to function continuously despite poor conditions and perpetual lack of maintenance. In fact, the durability of South African Rs (South African copy of the Israeli Galil), Russian AK/ANs, and the French FAMAS is well respected worldwide.

14 Jan 2003

Sage advice from a friend who finds himself overseas regularly:

"I was on assignment overseas late last year. My opinion on preferring a thirty-caliber (308) personal defensive rifle to one in 223 caliber was confirmed, in spades. I have an even stronger opinion on this subject than I did the last time we talked."

Comment: My friend has been doing this for some time. He is far more experienced than I would ever want to be. Every caliber has its place, but a thirty-caliber battle rifle surely covers all the bases!

16 Jan 2003

Comments on military rifle calibers from a major manufacturer of military small arms:

"As you noted, the 223 round (in any bullet weight) is effective only out to 150 meters. At 200 meters it is marginal. At 300 meters and beyond, it is impotent. Penetration is inadequate at any range. The 308 is surely an adequate caliber in both penetration and range, but there is a problem:

When armed with rifles chambered for 308, our people are invariably going to run out of ammunition before the enemy does, when the enemy is armed with AKs. Such inequities can be addressed with fire discipline, training, blah, blah, but most of us have come to the conclusion that a new military rifle cartridge is going to be necessary, one that will be effective out to 500 meters, penetrate like a 308, yet be of a size that will allow soldiers to carry quantities similar to the quantities of 223 they now carry. As I mentioned before, this project is in high gear at the Pentagon right now."

Comment: In the interim, we can fit most people into a 223 rifle. For light duty, like law enforcement and personal defense in a domestic environment (where ranges are short) it is adequate in the judgment of most of us. Rifles chambered for 308 are too big and heavy for small-statured people. But, when ranges open up, and one must shoot through things, only a 308 will do- at least at present.

28 Mar 2003

Black Rifles:

A friend e-mailed me today asking for a quotation on the subject of sighting-in "black" rifles. He is doing an article for a police magazine. So popular are they, he indicated, among police and non-police alike, that manufacturers can barely keep up with demand. These days, everyone wants one, it seems, even though you'll never hear about it from the news media.

Since 1994, many wonderful military rifles have been banned from import. Some states and localities have banned nearly all military rifles. However, most jurisdictions have not, and maybe it is time to look over what is available.

Serious calibers I recommend are 30-06, 308Win, 7.62X39Soviet, 30M1Carbine, and 223 Remington (5.56 NATO). Most black rifles these days are chambered for 223. I've editorialized many times on the inadequacy of the 223 as a military round, but for personal defense and domestic law enforcement, it is adequate, as is the 30M1 Carbine.

M1 Garand. M1s have not been banned from anywhere where you can still own any kind of gun. The rifle is big and clumsy, but what a warhorse! Chambered for 30-06, it is the biggest, heaviest, and most powerful individual rifle ever issued to troops, before or since. There are still plenty around, and they are easy to fix and work just fine. There is a learning curve on the reload, but, once one gets the knack of it, M1s reload plenty fast. Not a good choice for a small person, but still highly recommended.

1903 Springfield and 1917 American Enfield. Two great warhorses. Not autoloaders, but the bolt can be worked faster than most people think. As with the Garand, both can be owned nearly everywhere. The 30-06 round speaks with authority!

M1 Carbine. In my opinion, an ideal patrol rifle! Light, short, recoilless, has virtually no muzzle flash. Thirty years ago when I was a rookie patrolman, you saw M1 Carbines in many patrol cars. Today, they're coming back! They are easy to fix, as there is an abundance of spare parts and spare magazines. The 30M1 Carbine round is limited to one hundred meters, but, in most cases, there is plenty of power to stop fights. Like the Garand and the Springfield, M1 Carbines are unregulated in most places. Ideal rifle for a small-statued person.

Kalashnikov. Chambered for 7.62X39 Soviet and made from China to Bulgaria, the Kalashnikov Rifle has marched around the world, and for good reason. Rude and crude by western standards, the Kalashnikov is reliable and supremely usable. Imported ones are full of sharp corners and edges, and since 1994, they have not been imported. Neither has the Israeli version (Galil), nor the South African version, called the "R."

SKS. Also chambered for 7.62X39 Soviet, the SKS is also rude and crude but supremely reliable. Also banned from import, although there are still a lot around.

Happily, many wonderful rifles are currently being made in America and are available right now at retail gun shops.

DS Arms: Makes a wonderful FAL. It is my favorite serious rifle. Pricey, but worth it. If you want a 308, this is a great way to go. Normal capacity magazines (20 rnd) are still plentiful and cheap. DSA customer service is second to none.

Springfield Armory: Still makes their M14 copy, called the M1A. Unhappily, normal capacity magazines are scarce and expensive. Bushmaster, DPMS, Rock River Arms, and others: Make the famous AR-15/M4. Lots of spare parts and normal capacity magazines available. The rifle is hard to beat. If intended for serious use, the extractor needs a "D" ring.

Robinson Arms: Makes the wonderful RA-96, the most reliable 223 I own! It is my standard travel rifle, I am seldom without it. It takes standard, AR-

15 magazines.

Krebs Custom: Makers of custom Kalashnikovs. Sharp edges and corners are gone. Excellent western sights. Excellent trigger. Everything works smoothly. Nice package.

Every able-bodied American citizen of good character should own a military rifle and keep it in a reasonable state of readiness. With the international situation as exciting as it is, many, it seems, are heeding the call these days!

2025 Update: All the manufacturers mentioned above are still active and producing. Robinson Armament's flagship rifle is now the XCR.

29 Mar 2003

On the M1 Carbine from a friend on active duty overseas:

"A good friend (old at this point), one of Merrill's Marauders during WWII, dropped at lot of the Emperor's men with that 'little' cartridge. He was one of the first to carry the gun and fought all through Burma with it, the Carbine and a Colt Government Model. For a man who later became a renowned diplomat, he killed a lot of guys.

He used to coach us, and I saw him handle a Carbine many times. Even at his age, he could pump out eight rounds in the blink of an eye and make all hits in the neck and head of an IPSC target. He was faster than any of the rest of us. In addition, he was great at shooting on the move, always in a crouch, either forward or backing up. He said when with the Marauders, you were either chasing Japs or getting chased. I learned a lot from that old man!"

13 May 03

New Rifles from RA:

Alex Robinson of Robinson Arms in Salt Lake City informs me that they will be manufacturing two new rifles before the end of the year. Neither will be on a Kalashnikov, Garand, nor FAL platform.

One will be chambered for 7.62X39 Soviet and will use AK47 magazines. The other will be chambered for 308 and will use either FAL and/or M14 magazines.

These rifles, if they prove to be as reliable as the existing RA96, will be a great boon to mankind, and I'll report on them just as soon as I get my hands on one.

2025 Update: Robinson Armament is now a major American manufacturer of serious rifles, and their current flagship, the XCR, is second to none.

Recommended!

5 June 2003

New rifle for the USCG:

"Shortly, USCG drug interdiction/boarding teams will be armed with M-16s in an upper modified to chamber and shoot the 50AE cartridge. They will be using this weapon to disable the swift boats used by drug smugglers. The bullet has shown its ability to instantly shut down outboard motors and punch large holes in hulls. The ammunition is effective to 150 meters."

Comment: I'm sure much testing was done. I'm interested to see if the new rifle will really do what they claim. Testing is often agenda driven. Real-life results are more difficult to fake.

18 Aug 2003

More of Rifle Sights:

We did a Rifle Course here in Michigan last weekend. Here is a comment from one of our students:

"The reason I took so long shooting that course yesterday was because when I got to the line, I shouldered my AR normally but, but I had great

Rifle Advice

difficulty finding the correct eye-relief. After what seemed like an eternity, it occurred to me that I hadn't extended my collapsible stock, and my face was thus too %$#@!~ close to the %$#@!~ scope! I thought, 'Geeee, if I unshoulder and extend now, twenty people will see what a boob I am. Then, I remembered the words of my teacher, 'Don't whine about it. Just work through it' So I did..."

Lesson: Telescoping stocks on ARs, combined with close-eye-relief optics, can generate a potentially dangerous delay, as we see.

20 Aug 2003

More comments on equipment from several colleagues:

"I often see students here carry their CAR-15s with the stock fully collapsed (shortened). When in a deployment drill, they invariably forget to extend the stock and then subsequently drive their faces into the rear sight, or, like your student, fumble around trying to find the correct eye relief.

Our fix is to wrap a piece of electrical tape around the recoil spring tube just forward of the first or second notch extension. This maintains the correct index and prevents the shooter from collapsing the stock fully. I caution students not to carry the weapon on duty with the stock fully collapsed, as they will, without fail, experience the above problem at the worst time."

"Gun-writers and publishers are in bed with manufacturers, so every new attachment, 'enhancement,' and silly gadget receives glowing press in gun rags.

Humanity loses little when expensive and useless tinsel-toys fall off the weapons owned by naive gunshop commandos. It is another matter when people (who are mistakenly taken seriously) recommend things like forward slide serrations on autoloading pistols and then go on to teach techniques that lead to fingers being blown off. How difficult (or controversial) is it to know that the front of the pistol is where the bullet comes out, so one is well advised to keep his hands away from there?"

My comment: When we all were preparing for the delusory Y2K disaster, somebody asked me what they really needed. I said "a rifle that works and a canvas sack full of extra magazines and water." Of course, lacking glitter, glamor, and sex-appeal, that advice never made it into the press. What a surprise!

15 Sept 2003

Rifle Sights:

We just completed an Urban Rifle/Shotgun Course in Illinois. During our low-light exercise on Saturday evening, one of our students used an AR-15 with a three-dot, night-sight setup similar to what one finds on pistols. It was an abomination!

The two rear dots are too close to the shooter's face to do anything but compromise his night vision. The front dot was nearly impossible to see when it was superimposed on our steel targets (illuminated by road flares). This student, who shot well during the day, was unable to hit with any regularity at night. He is getting rid of the dots!

Another student used an ACOG optic. Again, the red crosshairs in the center of the reticle disappeared when superimposed on steel targets at night. The scope was mounted on the carrying handle of this student's AR-15, making a normal cheek weld impossible. He had an aftermarket cheek piece installed in order to solve this problem, but it continuously interfered with the normal operation of the charging handle. Again, this student's equipment was more a hindrance than a help.

Another student's rifle was equipped with a close-eye-relief, 4X scope. As we have come to expect, he continually got targets mixed up. He could usually hit; he just kept hitting the wrong target, because due to his narrow field of view, he got lost in his scope.

Students with conventional iron sights, red dots, and LPVOs did best, because their equipment didn't get in their way. They learned to develop a symbiotic synergy with their rifles instead of naively expecting that some attached miracle-gadget would magically substitute for personal competence and common sense.

17 Sept 2003

Muzzle Down!

In our defensive rifle and shotgun programs, I am emphasizing that students are well advised to keep the muzzle of their longarm angled downward, with that butt flat and above the shoulder as much of the time as possible. When mounted at eye level or in the depressed/ready position, I advise that the support hand we well forward on the forend, grasping it firmly, rather than merely resting the forend on an open hand. One is well advised to have strength on his weapon.

Retention is the reason. When an attacker can get within arm's reach and get under the barrel, pushing it up and toward the shooter, the shooter will find subsequently getting the weapon pointed at the attacker to be nearly impossible. He will probably have to default to his pistol, and fast! On the other hand, if an attacker can only grasp the barrel and forend from the top, the shooter can simply fall backward, the effect of which will be to get the weapon pointed at the attacker.

Retention is the forgotten imperative in much longarm training. Forget it at your peril! Much of the defensive longarm work that we do is at what would normally be considered pistol ranges. In fact, many times a pistol would be a superior weapon for such tasks. However, when one finds himself armed with a rifle or shotgun during such close-range encounters, he will probably not have the option of changing weapons. That being the case, retention becomes a crucial issue.

We must thus ask ourselves every time we train (with any weapon for that matter), "How retainable is my weapon when I'm in this posture or when I'm performing this technique?" Put another way, "When someone makes a serious attempt to disarm me right now, what will I do in response, and how successful will I likely be?"

Frail, impotent, and weak stances and postures may work fine during competitions, but remember, when participating in a quaint shooting contest, nobody will suddenly try to rip your weapon out of your hands and then shoot you with it. Next time you're confronting dangerous suspects, someone just might!

10 Nov 2003

From a Friend and Seasoned Rifleman:

"I shot my Springfield M1A last week. To my unhappiness, I discovered that my scout scope had lost its zero since the two of us were last together in Buena Vista, CO in July. Suddenly, it was shooting low and right. All the scope mountings were tight, but there was a crack in the composition stock, below the right side of the receiver. This doubtless reduced upward tension on the barrel and lowered the point of impact. I put the original walnut stock back on for now.

However, I decided to take the scope off. As low as it was mounted, it was still too high to get a satisfactory cheek weld. Few rifles (military or sporting) are set up for a good cheek weld with a scope. Even fewer have a secure way of mounting the scope on a rifle subject to rugged and heavy (military) use. Back to the old faithful iron sights!"

Lesson: Scopes, even scout scopes, will never be as rugged as iron sights. Every serious rifleman should be familiar with, and comfortable with, iron sights, and every serious rifle should be equipped with them.

19 Dec 2003

On the new military rifle caliber, from a friend close to the issue:

"The 5.56 NATO round is so entrenched, it is difficult to make changes. The Army knows full well it cannot do the job, yet they don't want to redo rifles. The Navy and Marines are Gung Ho for the new round, but they don't have enough pull by themselves. It looks to be a long time before anyone except special teams has it. Even special teams will be lucky to get it!"

Comment: There were indications that the new, 6.8mm round was on the fast track. Now, it is looking as if the whole project has stalled. "Nothing is too good for our men!"

16. Rifle Reliability

10 Dec 2002

"D" Ring

My colleague Jeff Chudwin brought the "D Ring" to the attention of all of us earlier this year. I have now, belatedly, joined him in his endorsement of this product.

All military rifles have issues, and the AR-15/M4 is no exception. One real weak point of the AR-15 system has been the extractor spring, which breaks at around 1,500 rounds. AR-15s with twenty-inch barrels that are fired only semi-auto are far less likely to experience this problem than are sixteen-inch rifles that have been exposed to heavy doses of full-auto fire. Even with the spring broken, the rifle will likely continue to fire normally. However, a broken extractor spring will invariably cause the extractor to release the fired case too soon, causing a live round to be stuck under a fired case that is still partially chambered. Called "soft extraction," this type of stoppage is particularly difficult to reduce and usually takes the rifle and the rifleman out of action for the better part of a minute, even when he knows what to do.

Until the bolt and extractor can be redesigned, the interim solution has been to place a rubber ring ("D Ring") over the extractor spring. Installation takes about a minute and can be easily done at the user level. The "D Ring" boosts the life of the extractor spring to 35,000 rounds and provides positive extraction and ejection even when the spring itself is broken. Seasoned operators worldwide have "D Rings" installed in their ARs. I have them installed in all my ARs.

All light military rifles, when subjected to heavy doses of full-auto fire, will experience extensive parts breakage. Pins will shake loose. Bolt lugs will crack. Springs will break. The rifle will literally shake itself to pieces. Your weapon was designed as a military, autoloading rifle and should be used as such. Using it as a "machine gun" exceeds its design specifications and brings with it all kinds of negative issues, as noted above. When I was an infantry officer in Vietnam, the full-auto feature was added to every M16 rifle, so that any rifle could be magically turned into an "automatic rifle," which was supposed to substitute for a machine gun. Unfortunately, on full auto, the M16 heated up so quickly that its usefulness as a "machine gun" was nil. True machine guns feature heavy, quick-change barrels. Without that feature, a light rifle that merely fires full auto is of little use, because it will quickly overheat. It cannot be counted upon for sustained fire. In practice, one is well advised to leave full-auto fire to legitimate machine guns and train himself to hit individual targets with carefully aimed, individual shots.

2025 Update: The D Ring is still advised on all AR/M4 rifles.

27 May 03

Of CETME rifles:

"At a recent Urban Rifle Course, two students brought Spanish CETME rifles. You see them advertised extensively. They are similar to the H&K G3 (91). However, their owners had to lubricate them constantly. They used both gun oil and bearing grease. As long as the guns were greased up heavily, they worked fine. As soon as the grease cooked out, they began malfunctioning."

Comment: We have the same situation with the H&K MP5. Heavily lubricated, they work best. Trouble begins when they dry out. It goes with the species.

10 July 2003

On weapons maintenance from a friend on active duty:

"On the release of the Army's 'official' report of the ambush of the 507th Maintenance CO in Iraq, I am motivated to comment on the sorry state of individual weapons maintenance training within the Army. I've heard many comments about design improvements that are necessary within the

Rifle Reliability

M-16 system, but they are out there in the future. In the interim, we need to make this system work.

Joe GI goes to the range and fires his 40-round qualifier, has a couple of stoppages (usually due to a broken part, poor lubrication, or a bad magazine), applies an immediate action drill (or takes an 'alibi') and logically concludes that this is all 'normal.' Well, it is not! Properly maintained, M-16s will fire hundreds of rounds consecutively without a hint of a stoppage. Broken parts and bad magazines need to be replaced on the spot, but they typically aren't, and substandard performance thus continues to be tolerated.

Soldiers need to be trained to maintain and lubricate weapons correctly, keep dust covers closed and magazines in the magazine well, and then not to accept individual weapons functioning poorly as 'normal.' "

Comment: Well said, my friend. I hope the right people are listening.

24 Aug 2003

We conducted a Defensive Urban Rifle Course here on the East Coast this weekend. In case I've neglected to do so thus far, let me list two items that don't work and should not be part of any professional gunman's inventory:

The Ruger Mini 30

40-round magazines for the AR-15

We've seen a number of Ruger Mini-30s in courses, and we had another this weekend. As we have come to expect, it was "malfunction junction" from start to finish! This rifle just doesn't work well, with any kind of magazine, nor with any species of ammunition. The student who owned it was completely frustrated and vowed early on to get rid of it as soon as he got home. I concurred!

Another student had a Bushmaster AR-15, but he was using two after-market, 40-round magazines. Neither worked. Failures to feed were rampant, particularly involving the last few rounds. Next day, we switched him over to 30-round magazines, and problems instantly disappeared. The rifle functioned perfectly from that point forward. The two 40-rounders ended their unhappy careers in the trash barrel!

Lesson: The training range is the best place to discover flaws in technique and equipment. These two students had the courage to face facts squarely and make the corrections necessary. They're going to be okay!

26 Aug 2003

More rifle comments from another colleague:

"In Vietnam nearly forty years ago, I carried a Colt Car-15, as did you. After several days on patrol in the rain, and not having made enemy contact, our colonel decided that we should be picked up and reinserted in a different area. When I entered the helicopter, I attempted to push the manual safety to the 'on' position, only to discover that the safety lever was frozen solid. Pounding with a knife butt finally caused the lever to come loose.

Back at base camp, inspection revealed that detent holes in the safety body had filled with water and had subsequently rusted shut. The cure was, of course, to clean and lube, then apply grease, filling the holes and groove completely. The problem never again reared its ugly head.

Never, that is until decades later when I was going through a course at Blackwater during a rainy week, again with a Car-15, this one manufactured by Bushmaster. Once again, at the end of the day, I could not move the safety lever. Inspection revealed the same problem!"

Lesson: The above-mentioned incidents could have occurred with any rifle. During active operations, weapons must be function-checked constantly. During long periods of inaction, continuous inspection is even more important. Rust and corrosion affect any species of metal.

The unexpected incessantly stalks the unwary!

17. Shotgun Advice

25 Mar 2003

This is from a friend on the subject of the defensive shotgun. Excellent stuff:

"What rate of buckshot dispersion is optimal? How fast is too fast (pellets off target)? How slow is too slow?

The prime missions of the shotgun are:

To defend against any EXPECTED, close-range attack, and

To defend against moving targets (particularly in low light), within twenty-five meters.

I contend some dispersion is desirable. An excessively small pattern requires the same precision that I would need with a slug. It gives up too much flexibility when the target is moving and I am also moving. I like the classic '2cm/meter of range' dispersion rate of the typical 'Police Cylinder' shotgun choke. It renders a nominal 20cm diameter pattern at ten meters.

At a class last summer, a student was using a 'special' barrel and low-recoil 00 buckshot. At eight meters his 'pattern' was one, ragged hole. Too small for our purposes. By contrast, my pattern at the same range was 15cm (6"). Add movement of both your adversary and you, and you'll appreciate the advantage of this dispersion rate.

For precision shots (at any range) or shots at a distance greater than twenty meters, I will opt for a slug anyway.

A screw-in muzzle-device, called the 'Wad Wizard' is becoming popular. Little 'feet' catch and retard the plastic wad, so that it does not blow through (and subsequently be overtaken by) the shot pattern in flight. It gets rid of 'flyer pellets' and greatly contributes to the homogeneity of the pattern.

Barrel porting is for a "game guns." Close-in weapon retention firing techniques may require the support hand to grasp the barrel just proximal to the front sight. Porting will expel gasses near the support-hand and upward in the direction of the shooter's face. In my opinion, there is no place for porting on ANY defensive firearm.

I love shotguns (almost as much as rifles). Shotguns are optimal in the 5–20-meter range and (with slugs) can be pressed into service as a substitute rifle at ranges out to seventy-five meters."

Comment: He lays it out well. As a fight stopper, shotguns have no equal!

2025 Update: "Flight-right" ammunition technology, with its high-drag wad, has made the Wad Wizard largely unnecessary. However, I still recommend conventional, non-magnum, standard-length, nine-pellet, 00 buckshot rounds for serious self-defense. I don't recommend any "low-recoil" shotgun ammunition.

27 Mar 2003

Shotgun sighting:

A student at a recent Defensive Shotgun Course used a Remington 870 in 20ga with a standard, bead front sight and no rear sight. Her technique was good, but she had difficulty striking rotating steel targets with enough force to cause them to rotate. She was using WW 2 3/4" #3Bk. Range was eight meters. I could see that she was hitting the target, but it was barely moving. I asked to shoot her gun myself, and I easily rotated the target with only two hits. It then (finally) dawned on me what was happening:

Her earmuffs were preventing her from acquiring a correct cheek weld (by forcing her head too far up). In addition, the stock was too long. The combination of circumstances caused her to shoot high, as her eye was too high over the barrel. Even though she put the front bead on the target, she was consistently hitting the top edge of the target with the bottom edge of her pattern.

One of my instructors saw the problem and graciously lent her a set of slim

Shotgun Advice

earmuffs. She was then able to get a correct cheek weld (even though the stock was still too long), and she immediately starting rotating targets with the rest of us.

Had her shotgun been equipped with ghost ring or Express sights, an RDO (red-dot optic) or standard pistol sights, she probably would have been able to see and solve the problem herself much earlier.

Lessons: (1) Most shotgun stocks are too long for even an average-sized male. All are too long for the vast majority of females. Women's shotgun (and rifle) stocks nearly always have to be shortened (sometimes by as much as three inches) in order for the shooter to be able to acquire a correct cheek weld.

(2) Defensive shotguns work best when equipped with Express, ghost-rings, or RDOs. A front bead (with no rear sight) is surely usable, but it is not the best setup. When you have a shotgun with a bead front sight and no rear sight, as the weapon is mounted, if you can see the barrel at all, you're going to shoot high. You should see the crown of the receiver and the bead sitting on top of it. You should not see the top of the barrel (or rib) at all.

(3) With the stock length adjusted properly and the gun equipped with reasonable sights, all female shooters I've seen are deadly with a shotgun! Twelve-gauge shotguns are unpleasant for shooters weighing less than 160lbs. However, we have had great success with 20ga shotguns in the hands of most of our petite female shooters.

3 Dec 2003

Mossberg M590A1

I have new Mossberg 590A1 pump shotgun (12ga) that I have been using for the last several weeks. It makes a nice package!

It came with an 18" barrel, speed-feed stock (four extra rounds in the stock itself), and large, rugged, ghost ring sights. Rear sight is well protected, and the front sight is beefy and easy to see. Magazine tube holds five rounds. I surely like the Remington 870 also, but Remington doesn't offer such a complete package from the factory.

There is a slight learning curve with the speed-feed stock, but it is plenty fast and extremely convenient. It doesn't modify the gun's outline or create a snag as does a sidesaddle.

In some states and jurisdictions, military rifles are aggressively regulated. In most of those same areas, shotguns are essentially unregulated. Indeed, the demand for Urban Rifle training in those areas is low (outside the law enforcement community), but the demand for defensive shotgun training remains high, and probably will for the foreseeable future.

As a patrol shotgun, it is hard to beat. Also makes a nice "bedroom gun." Highly recommended!

5 Dec 2003

Comments on the Mossberg 590A1 from users:

"We have several Mossberg 590s here at the police range with over 100,000 rounds through them, mostly slugs. They have all had minimal maintenance and are still perfectly serviceable.

Some important notes on the Mossberg:

For some reason, the guns are shipped from the factory with a coating of grease on the internal fire control mechanism. This sometimes causes sluggish operation. A detailed disassembly of new guns is recommended to remove the grease and restore full reliability. This degree of disassembly should only be performed an armorer. It's standard procedure with new 590s here.

The red insert on the front sight is sometimes loose. We've had and a number of them fall out. When it gets loose, it needs to be re-staked and epoxyed back in place. Of course, without it, the gun is still functional. I don't consider colored sight inserts particularly important.

The springs in speed-feed stocks will sometimes take a set, causing the second shell to fail to come all the way forward. One can still get it out. It's just sluggish. I usually just stretch the spring, and that lasts six months or so.

The 590's manual safety lever is plastic and breaks now and then. Brownells makes a metal replacement, and I recommend it.

By the way, the Mossberg is the only shotgun that has a military-specification manual safety. With the manual safety "on," the gun is genuinely drop safe. No so with most other shotguns."

8 Dec 2003

More Mossberg 590 comments from a friend on active duty:

"We used 590s in Panama and they proved themselves both reliable and durable. We put 10,000 rounds per year through each of them with minimal armorer maintenance. Concur with your advice on replacing the plastic manual safety lever. It doesn't last long under field conditions.

When I arrived in Panama and took command of a security platoon. Two of our site posts used 590s, modified with after-market pistol grips and no shoulder stock (about as useless a contraption as can be imagined!). Upon touring the posts the first day I asked the Marines if they had ever trained with shotguns in that configuration. To my surprise, they had not.

I insisted that the pistol-grip-equipped shotguns brought to the range for our weekly 'port and starboard' firearms training. Sure enough, the Marines had little problem handling the shotguns when shooting light (birdshot), training rounds. However, when 00 Buck and slugs were fired, the guns were utterly uncontrollable and completely inaccurate, no matter how they were held. The pistol grips all promptly cracked as well.

My Marines quietly approached me afterward and asked that the pistol grips be discarded and that factory stocks be reinstalled. That very day, they were!"

Lesson: Emergency equipment should never be fielded without adequate testing. What looks sexy and shiny in the showroom often disappoints in the field. How many times we've seen it!

18. After Market Parts & Other Gear

3 Mar 2003

1911s

We just completed a defensive pistol class in Oklahoma City, OK. OKC is "cocked-and-locked country," as is much of Oklahoma and Texas. Most students had Kimber 1911s, and all worked well. Two had Glocks.

What didn't work well was the eight-round magazines used by many of the 1911 shooters. There were several different brands, and none worked reliably. Slide lock failure and failure to feed were seen time and again.

The Colt/Browning 1911 pistol (45ACP) was, at the outset, designed to accept a seven-round magazine, and I strongly recommend serious users stick with the original. They've worked well for decades. There is no brand of eight-round magazine I recommend for serious use.

Another excellent way to deadline a 1911 is through the use of extended magazines. They invariably insert too far (because they can) and cause a terminal stoppage. Also not recommended and, in my classes, not allowed.

5 May 03

Of XDs and magazine funnels:

We just completed a defensive pistol course in Sioux Falls, SD. One of our students works at a local gun shop and brought a SA XD pistol (compact) in 9mm. We'll seen a lot of XDs in classes, and they've all worked well.

This is the first compact model I've seen. It worked well, and now I really like the size. It will make a suitable concealed carry gun. Of course, normal capacity magazines are not available, but, with this compact model, one doesn't give up much capacity, since the magazine wouldn't be able to hold many more than ten rounds anyway. The compact model is now available only in 9mm, but a 40S&W version will surely follow shortly.

XD pistols "feature" a grip safety. Unhappily, unless fully depressed, the grip safety prevents the slide from moving. It is a confounded nuisance during loading and unloading, particularly for students with small hands. It is not a deal-buster, but I surely wish it were not there.

Several female students remarked that the XD trigger reach is considerably less than on the Glock. I had not fully appreciated this before, but they were right. It is thus a good choice for those with small hands, since the trigger can be accessed normally without having to compromise the master grip (often necessary with Glocks). In addition, "finger drag" on the pistol's frame is eliminated.

XDs are also less expensive than Glocks, and I believe this compact model will thus provide Glock with significant competition in the carry-gun market.

In our Class, we also had several 1911s equipped with add-on external aftermarket magazine funnels. I wish I had a dime for every failed reload caused directly by this abominable gadget. Time after time, these students would complete a reload, only to discover that they had chambered thin air instead of a live round!

I have scant regard for most aftermarket add-ons, but this ill-conceived contrivance is particularly vexatious. It adds bulk and weight as well as several exposed sharp edges, all in addition to promoting failed reloads. Not recommended for any serious pistol.

11 June 2003

Notes from the IALEFI Conference in Orlando, FL:

S&W had a strong presence at the vendor's show today and last night.

The SIGMA pistol is still in the product line and will remain. It is now their "low-cost" pistol.

After Market Parts & Other Gear

The P99 is now offered in the "Revolver Trigger" mode, the "Glock Trigger" mode (called the "QA"), and the "Manual Decocking" mode. The "Glock Trigger" mode is nice, but, on all models, the trigger still starts too far forward for those with small hands.

They tell me that some departments who have purchased the manual-decocking version of the P99 are now making the act of manual decocking optional! Officers are given the option of never decocking. That moves the trigger into a permanent, rear position, solving the trigger-too-far-forward problem, but, in my opinion, not giving the officer sufficient control over the trigger. Time will tell.

The P99 system is now available in a single column 45ACP. Grip and pistol are comfortable. Nice gun in that caliber.

A magazine safety does not come with the P99 but is available as an option. Most are already familiar with my opinion of magazine safeties.

Because of its light weight, recoil is sharp on the P99 in 40S&W.

S&W's 1911 is a nice gun. Feeds Cor-Bon PowerBall like hardball. I'm going to have to get one!

I finally had the opportunity to get my hands on an "X-Frame" revolver in 500 S&W and shoot it. A monstrosity (albeit an elegant one)!

I also had the opportunity to use the new SIG self-decocking trigger system (mounted on a 229). It has no name yet, but we all liked it. Pull weight is a consistent, smooth, seven pounds and short. Hammer spur is gone. It can be used in conjunction with either a regular or a short trigger. Vast improvement over SIG's existing self-decocking ("flat revolver") system. I'll have one soon.

Cor-Bon PowerBall in 45ACP, 40S&W, and 9mm fed flawlessly in all pistols present. This is the best round to use in just about any pistol. High speed, unimpaired functioning, adequate penetration, consistent and spectacular expansion, no plugging of the hollow cavity with clothing, benign appearance. Nothing not to like!

FN is now making the Hi-Power in the "SFS" trigger system. In this system, after a round is chambered, the operator manually pushes the hammer spur forward. The hammer spur actually disconnects from the hammer itself, so the gun is still "cocked." It just doesn't look that way. When one needs to shoot, pushing the manual safety down "recocks" the hammer (reconnects the hammer spur to the hammer itself). It a clever, cosmetic feature, designed to mollify those uninformed few who might see a cocked hammer and perceive a hazard. Friend, Bill Laughridge used to offer this system as an option on Hi-Powers. Now, FN offers it from the factory. The downside is that it always takes two hands to reholster.

The Beretta (pistol cartridge) rifle, called the CX4 (Storm), is streamlined, comfortable, and handy. Unfortunately, the crossbolt manual safety is misplaced and impossible to use without compromising one's grasp on the gun, and the forend gets too hot to grasp under rapid fire. The rifle only takes Beretta pistol magazines, another mistake. Ruger made the same mistake with their pistol-caliber rifles, by only making them to accept Ruger pistol magazines. If both the Beretta CX4 and the Ruger rifles accepted Glock magazines, they couldn't make them fast enough!

Remington is now marketing their M7600 pump rifle as a police carbine. The caliber on display was 308. It will have a hard time competing with the DSA/FAL, but it is cheap by comparison.

An Israeli company is producing the "Corner Shot," which is an elaborate, self-contained device that holds a Glock pistol in the front and enables the operator to look at a television monitor in the rear as he peers around corners. Clever, but it has a sizable learning curve and it is expensive. In addition, it is a real hazard on the range, since the operator can inadvertently point the muzzle sideways before he knows it, and the sideways recoil is a strange sensation! I witnessed several occasions where students inadvertently pointed the gun sideways down the firing line!

Beamhit, represented by the always exuberant (and my good friend) Brian Felter, has greatly expanded their product line. They are now competing with more expensive video simulators. It is a clever system on which many mechanical skills can be trained in, all in a non-shooting environment.

Snail Systems displayed their "flowing water" bullet trap. It's a conventional, steel escalator trap with a constant, downward flow of water over the impact area. The effect is that incoming bullets never actually hit the steel. Instead, they hit the water and hydroplane upward and into the snail chamber for final de-energization. Dust is eliminated, as is noise and wear and tear on the steel itself. For a heavy use range, it is a great idea.

Advanced Training Systems is now making dense, rubber sheeting for range walls and, in some cases, impact areas. Made from recycled tires, it is inexpensive and long lasting. It can even be used for knockdown targets.

Lasermax had a nice display. The unit fits completely inside a Glock, SIG, 1911, and Beretta. I used it in conjunction with a partner (who also had one). I found that the laser can actually be a nonverbal form of communication with a partner. If one of us lasers a suspect, the partner can see and knows to laser the second suspect. I may have to soften my opinion of lasers on pistols.

The Prism video simulator was on display. Excellent but expensive. The system is now available in an IMAX format, with screens on the sides and back, forcing the student to look all around.

I used a rifle with an EO Tech sight on several Prism scenarios. I dialed down the brightness of the illuminated aiming point, perceiving it as too bright. However, seconds later, when I was under stress, it became obvious that the aiming point was not bright enough! Next time, I'll know to make it brighter than I think necessary.

I noticed that the EO Tech System encouraged me to shoot too fast. I had to make myself slow down.

More later.

8 Aug 2003

TiAlN Coating

"I just had a 1911 treated with TiAlN (titanium aluminum nitride) by Molecular Metallurgy here in CA. My Colt is now a uniform, dark matte charcoal gray, almost black. Job was great, but I ordered a matte finish on all parts.

I should have specified a polished finish on the barrel throat and feed ramp. When I first took the old Colt to the range to test fire it, it wouldn't feed anything, not even hardball! Rounds kept hanging up on the matte finished feed ramp.

When I got it home, I tried to clean up the feed ramp with a Dremel buffing wheel and some polishing grit but couldn't make any progress. It seems about the only thing that will scratch TiAlN is more of the same. So now my final cleaning procedure for the Colt is to manually cycle half a dozen or so hardball rounds through it to 'lube' the feed ramp. As long as I do that, it remains reliable."

Comment: When you have something done to a serious gun, but sure that the people doing the work know what the gun is for and that your life may depend on it. All serious guns need to be thoroughly tested before being placed into service. The world is full of surprises!

8 Sept 2003

From a friend who just attended a week-long pistol course at a well-known school:

"We shot 1000-1200 rounds during the course of the week. Most students brought 1911s. All of them had 'customizing' and aftermarket parts. All 1911s experienced feeding and ejection problems, some chronic. Several broke parts and went down after the first two days.

There was a single Taurus pistol. It fell apart the first day.

A SIG 220 ran well all week, as did a G21. Both were 'out of the box.'

This school likes 1911s, but the performance of the ones in this class was less than inspiring, particularly those that had been 'customized.'"

Lesson: The 1911 system is inherently sound, but its devotees will not be dissuaded from tinkering with it. Usually a bad idea, as we see.

19. Hunting

18 Nov 2002

Our annual pig hunt in Ohio is now history. We had a wonderful time, as always.

Twenty-five years ago, my father gave me a Colt/Sauer "Grand African" bolt-action rifle, in 458 Win Mg. It is equipped with express sights. I remember shooting two or three rounds through it way back then. It has been sitting in my safe ever since.

So, I decided to use it on the pig hunt this year. We need to put all our equipment to practical use, said I. No shrines for me!

I shot two pigs with it, one at twenty meters in heavy brush, and one at thirty meters in the open. I used 510gr soft points. Both shots were offhand. Both pigs were hit from the side on point of the shoulder (through and through), and both went down immediately.

When I sighted it in the afternoon before the hunt, I was intimidated. Recoil was everything you would imagine, and I wondered (silently) if I could shoot it accurately at a real target.

As it turned out, when I shot the pigs the following day, I had no recollection of the recoil at all. My fears, like most fears, were unjustified.

I learned, once more, that it is just a rifle, and I can shoot it accurately, just like any other rifle. The best way to deal with fears of all kind is to confront them directly. My first chief of police once gave me this sage advice,

"Always call their bluff."

He was right!

23 Feb 2003

I'm back in Colorado from my 2003 Africa trip, safe and sound! Details to follow. We did several classes. One of our students was the head of South

Africa's "FBI" (called "Skoprions"). May be back to train the rest of them.

I shot four fine animals, including an eland and a black wildebeest. Should have been five, but for a muffed shot.

I'll be catching up on e-mail today.

6 Nov 2003

Report from the 2003 Pig Hunt in OH:

We just completed our annual pig hunt. We all gathered at a hunting preserve in southern Ohio. We took nine hogs this year, ranging in weight from 150 to 400lbs. There were nine of us. We were all successful, to one degree or another. There was one Benelli Super-90 shooting Foster slugs, one 44Mg Ruger lever gun, one 6mm bolt gun, one 303Br SMLE Enfield, one 308 bolt gun w/scout scope, one 30-06 bolt gun, one G20 10mm pistol w/six-inch barrel, one 45-70 lever gun w/Cor-Bon 350gr SP, and I used a Chinese Kalashnikov in 7.62X39 (30 Soviet). I didn't personally witness all shots that were fired, but I saw most.

This kind of hunting is fast. Targets come and go quickly and always seem to appear when not expected. Windows are typically no more than two seconds. Hunting terrain is such that ranges are ten to a maximum of fifty meters. Pigs are dense, tough, amazingly quick, and close to the ground. They don't usually "topple" when hit, and multiple hits are difficult, even with an autoloader. Your first hit better be good!

Observations:

Iron sights work well, but scout scopes (LPVO, Low-Power Variable Optic) work well too! A forward-mounted, low-profile, One to four-times magnification scope is well suited for pig hunting. By contrast, high magnification, close-eye-relief scopes are counterproductive, indeed close to useless. Most of us used iron sights.

In profile shots, the best place to aim is the point of the shoulder. Nearly all "instant" stops involved hits here. A bullet passing through this point

Hunting

breaks the shoulders and jars the spine. When thus struck, most pigs drop on the spot or run only a short distance. By contrast, gut shots, from any caliber, will guarantee a long chase and maybe a lost animal. I've never seen a gut shot put an animal down quickly. Heart/lung shots are effective, eventually, but the animal nearly always runs afterward, sometimes a good distance. The 6mm shooter hit a pig just behind the shoulder (fifty meters). The 100gr SP bullet went through and through, and the pig left a substantial blood trail, but his legs still worked, and he ran and ran!

The 10mm pistol round was disappointing. The shooter had a good hit (which I personally witnessed). The bullet struck from the front and went into the right shoulder, just to the side of the jaw. It was a 180gr WW Black Talon. I could see the pig was hit, but he still ran and ran. He was obviously uncomfortable, but he kept going. The shooter finally (after five minutes) had to borrow the Ruger 44Mg Carbine to ultimately stop the flight. Inadequate penetration was the issue with the 10mm.

Instant stops were registered with the 12ga slug, the 45-70, the 44mg carbine, and the 308. Good calibers and competent shooters. All went through and through, except the 44Mg. I was presented with an angle-away shot at twenty meters. My first bullet hit just behind the rib cage and ranged forward (125gr Federal SP, no exit). I caught the link and tried to get a second hit, but the pig was gone. He ran fifty meters before collapsing.

I like this kind of hunting, because it gives me the opportunity to fellowship with my friends, the opportunity to use military weapons, but it also reminds me that I need to be observant, reactive, and decisive, and that I won't get a second chance to be competent.

Fooling around with inadequate calibers in an effort to "push the envelope" is something that doesn't interest me, and it interests me less every time I see it. A bullet needs velocity, mass, penetration, expansion, and structural integrity (doesn't come apart). Inadequacy in any category negates the rest.

Ask me how I know this!

27 Dec 2003

Important lesson from Bob Ruark in Use Enough Gun:

"... I put the bead somewhere on his neck behind the ear and squeezed. The big Wesley-Richards, which I trusted so much, roared and possibly kicked, but I never felt it.

The tiger never left his crouch over the dead buffalo. He never moved his head. His chin dropped an inch and came to rest on the buffalo's flank. He did not flex his forearms. He did not kick. He was stone-dead on the body of his victim; his eyes closed in the strong light of the torch.

I raised the gun again to give him the other half, the finisher, the tenderizer.

'No, sahib,' Khan Sahib said. 'Don't shoot again. He is dead... Do not spoil the hide. Nobody ever killed a tiger any more dead.'

... Kahn Sahib shook me by the hand and beat me on the back and danced up and down... I took the little flask of emergency ointment and had a long pull at it. I toasted the tiger. I toasted Kahn Sahib, and I toasted me... One-shot Bob. Some people miss 'em. Some people wound 'em. Not the Boy Genius. He shoots them in the neck.

I had just run for president and been elected. I had just been reelected to a second term and had won the water-boiling contest at the Campfire Club when I heard an awful roar. Kahn... flicked on the light just in time to see my dead tiger's tail disappear.

... the tiger had got up and gone away. I knew right then that I would never see that tiger again, although all common sense told me that this was a death flurry... We looked him high , and we looked him low for two days... No tiger. One-Bullet Bob... the new president of the jerk factory.

I could hear Harry's English Schoolboy voice saying, 'When it's big, and it's dangerous, shoot it once and shoot it twice, and, when you're absolutely certain it's dead, shoot it again. It's the dead ones that get up and kill you.'"

Comment: Bob Ruark died in London in 1965. He hunted during the Golden Days, and his lessons, which he shares with us in his own inimitable style, are timeless.

My experience with the oryx in the Karoo Desert in South Africa in 1999 was similar, and I'm lucky the beast chose to run away from us rather than at us. My oryx dropped with a single shot from my borrowed 270. He dropped as if he had been struck by lightning. Nary a flinch nor a shudder.

Like Ruark, I thought I was hot stuff and had splendidly succeeded in impressing my African hosts. Unfortunately, I had failed to impress the oryx, who miraculously sprang back to life twenty minutes later, long after my guide had insisted on unloading the 270. As I watched him casually lope away, displaying little discomfort, I, too, felt like the new president of the jerk factory!

Like Bob, I learned my lesson that day. In retrospect, it was a cheap one. Only my inflated ego was wounded.

"Experience never errors; what alone may error is our judgment."

Leonardo da Vinci

29 Dec 2003

Sage comments from Joe DaSilva, my friend and guide in Africa. Joe has kept me from being injured more than once. I owe him a lot. He knows what he is talking about:

"I have been hunting and guiding in Africa for thirty years, and I need to comment on your Ruark posting:

Firstly, the biggest mistake first-time African hunters make is failing to study the anatomy and habits of our game animals here. Time and effort in reading and speaking with people who know game in Africa will pay big dividends. Hunting here is serious business. Hunters are grievously injured here every year. Some are killed. Many more are unsuccessful and frustrated. The more you know, the luckier you get!

Secondly, correctly choosing the rifle, caliber, mounts, and scope for the type of animals that you are going to hunt. Again, you need to listen to experienced people and take their advice seriously. Over the years I've seen many guys and gals with ill-selected equipment either (1) lose game, or (2) damage the animal so badly that the taxidermist is unable to create a suitable mount.

Thirdly, you need to study the terrain where you will be hunting. For example, if you are hunting antelope in heavy bush, you will be shooting at ranges that will not exceed one hundred meters. You may have shots as close as twenty meters! For this, you don't need a 7mmRmag. A 270, 308, or 30-06 will do the job very well.

Finally, most hunters who come to Africa are not accomplished riflemen and have not practiced nearly enough. Worse yet, they don't want to listen to advice from those of us who live here. They invariably bring rifles in an inappropriate caliber. The routine result is that they become gun shy and miss their shots. They miss consistently, because they have created a huge flinch, a poor and inconsistent cheek weld, and are thus scared to take a second shot. I see many guys with half-moon (scope) cuts above their eye, and, oh boy, do they bleed! After such an experience, most hunters are finished for the day, maybe even their entire trip.

The two rifles we like most here are the 300H&H and the 458WinMag. Doubles are glamorous, but I like bolt guns. I have killed twenty-four cape buffalo and, in all that time, have been charged twice. I can tell you confidently that the 458WM is more than enough. Controlled, accurate shooting is the key! You cannot allow yourself to panic. Understand that you'll have to hit the beast more than once. Most guys with 378s, 470s, 450s, and 500s cannot even make a viable second shot because of recoil. As a result, the wounded animal trots away, or charges! What's the use of having it?"

20. Duty Gear

18 Oct 2002

Latest from LAPD:

"Our new Chief, Bratton, who brought Glocks into Philadelphia and NYC, is now bringing them into the LAPD! I suspect we'll finally see our Berettas go. Our SWAT folks like Glocks but are sticking with the 1911.

Bratton has already started to clean house with our command staff. He wants resumes from everyone above the rank of captain. Many have already announced their retirement!

All in all, he looks like a breath of fresh air to me. Heaven knows, this department needs it!"

25 Feb 2003

Cold Steel in Africa

The most popular serious knifes in Africa today are, as you might expect, those made by my friend Lynn Thompson at Cold Steel. In terms of functional design, strength, and sharpness, they are second to none. Unfortunately, the current exchange rate makes them prohibitively expensive for many Africans. All my friends have Cold Steel knives and are extremely grateful that they do.

By contrast, most locally made knives a cheap junk, but they are found everywhere and are still extremely dangerous in the hands of evil doers. There are thousands of knife homicides every year in South Africa. Tens of thousands of disabling and disfiguring injuries. Subsequent infection kills many more. Most never find their way to statistics that you will ever see. The preferred method of attack is multiple downward (icepick) stabs into the shoulder (from the front or rear), the target being the subclavian artery. It is effective and difficult to see coming.

As one might expect, the local knife culture is experienced and active. One of my knife instructors from the local AMOK School in Capetown made a good point during one of our recent sessions there:

When there is a large amount of blunt trauma associated with a cut, the body has the opportunity to restrict the flow of blood to the area and thus limit bleeding. By contrast, when there is little trauma associated with a cut, bleeding is always maximized. The implication is clear: when using a knife in self-defense, the sharper the better. Sharp knives will always produce more rapid bleeding than will dull ones. As you probably know, Cold Steel knives come to us "dead sharp." I don't know how they get them so sharp, but their ability to cause catastrophic hemorrhage is unexcelled.

The lesson here is: Don't use your serious knife for utility cutting. It will rapidly become dull and then serve you poorly when you really need it. Keep your fighting knifes dead sharp. For utility cutting, use a utility knife that is not intended for serious purposes.

"No compunction
Just reactions
No regrets
No retractions
One with my blade, alive in my hand
Pure of heart and firm in my stand
It's not in the blade; it's played in the mind
Seek the True Way,
and the Way you shall find."

8 June 2003

From a friend with the LAPD:

"The rumors that have relentlessly circulated since Bratton became our new chief several months ago are coming true! The Beretta 92F that we have used here for so many years will be phased out in favor of the Glock pistol, probably the G22 and/or G23. I, for one, am glad to see the caliber upgrade.

A number of reasons have been cited. Our existing fleet of pistols is due (actually overdue) for retirement, and an upgrade to 40S&W has been in the works for some time. However, durability issues with the Beretta 96F have effectively eliminated it from consideration.

All our firearms instructors are purchasing personal G23s, so they can be ready for the switch."

Comment: The LAPD is the last, big department to give up on the Beretta pistol. Beretta, once king of the hill, has steadily lost market share to the point where today they can be called "non-players."

29 June 2003

Latest from LAPD:

"Our Police Commission is about to approve the Glock pistol for the LAPD. It is a forgone conclusion. The chief wants them. We'll have 9mm, .40S&W, and 45 ACP. Officer's choice. Our department firearms instructors have already gone through a Glock transition course.

The official round will be Winchester/Ranger.

Officers will have the option of retaining their Beretta 92Fs, but the majority are expected to make the switch."

Comment: It will be interesting to see how many make the switch and how fast.

2025 Update: Berettas were all gone within sixty days!

3 July 2003

It official at the LAPD:

"Below is the column Chief Bratton writes in our Police Protective League's monthly paper, The Thin Blue Line. It arrived in my mail box a few minutes ago:

'New Glock Weapons

We are currently preparing correspondence seeking the Police Commission's approval for officers to carry Glock 9mm, 40 caliber, and 45 caliber firearms. This will be the first time that the Department has authorized the use of Glocks. Although the City won't issue Glocks, officers will be able to purchase these firearms on their own and carry them on duty after receiving department training.'

When the chief puts it in The Thin Blue Line, it is as official as it gets."

17 July 2003

More from SIG:

Law-enforcement orders for pistols in 9mm have dwindled to a trickle. In fact, the 228, once SIG's flagship, is now all but out of production. The only big department that still issues it is the NJSP.

In addition, pistols chambered for 45ACP are no more popular than those chambered for 9mm. Although the popularity of the 45ACP does not seem to be diminishing at the same rate as that of the 9mm, it is surely not increasing either. Most 45ACP fans end up with a 1911 clone from Kimber, SA, or S&W.

The vast majority of police orders are for pistols chambered for 40S&W, although the 357SIG is now coming on strong and indeed starting to pull up alongside the popularity of the 40. Those two calibers are what everyone wants now. One big reason is that these pistols can be easily converted from one caliber to another, a big attraction for chiefs of police who are trying to make all their officers happy.

Neither the 9mm nor the 45ACP is in any danger of dying out, but both have surely been pushed to the side!

>2025 Update: During the subsequent two decades, the 40S&W and 357SIG have both fallen into dispopularity, and the 9mm is back! Pistols chambered for 40S&W and 357SIG are loud, hard-recoiling, and

unpleasant to shoot. Ammunition is expensive and wear and tear on pistols is far more severe than with the 9mm. The 40S&W and 357SIG enjoyed popularity for a short time, but both have now fallen by the wayside.

23 Sept 2003

M9 pistol (Beretta 92F)

There is much stirring recently within the USMC and several other serious military units with regard to the 9mm M9 pistol, which displaced the 45ACP 1911 pistol back in 1985. In the intervening eighteen years, the M9 pistol has gained the dubious title of the most disliked, most disparaged, most unpopular firearm ever issued to American soldiers since the discredited French CSRG (Chauchat) light machine gun was foisted upon American soldiers in WWI.

I remember the JSAP project that selected the M9. The M9 was supposed to supersede all other pistols, so there would be only one pistol in the entire military system. The move to 9X19mm caliber was preordained, since that is the pistol caliber that was (and still is) used by our NATO allies. The fact that US soldiers had a separate pistol caliber was never an issue during WWII, Korea, or Vietnam. But, by 1985 this discrepancy apparently became intolerable, at least in the minds of "one-world" politicians.

The cover story was that the Beretta pistol outperformed all others during tests. The real reason was, of course, the Italians wanted to sell us expensive guns and also get a manufacturing foothold on US soil. So, they threatened to close our air bases in Italy if we didn't buy their overpriced pistols. In other words, the "selection process" was politically influenced, as it is for most other military equipment.

In any event, eighteen years later, the M9 has shown itself to be reliable, but not durable. Parts breakage is the big problem. Thus, keeping them running is the real issue, particularly when they are fired a lot. The pistol is big, fat, bulky, and not suitable for concealed carry, as is often a requirement these days, even for soldiers. The system for operating the two-stage decocking lever is surely learnable, but, as we all know, the Army is afraid of guns, indeed petrified of loaded guns, so soldiers (even officers) are commanded never to load the pistol anyway. Smart ones ignore such stupid orders, get competent training (outside the military), and then carry their pistols loaded anyway, as I have reported.

So legion have been the complaints about the M9, that SIG pistols have now made their way into the system, and even 1911 pistols (in 45ACP) can be seen in the hands of certain elite units that have enough political autonomy to ignore orders to use only issue gear. We're now back to multiple pistols, the exact thing the JSAP project was designed to end forever.

Suggested solutions range from dumping pistols altogether and issuing short rifles instead, to returning to the 1911. Neither extreme is likely to see the light of day. Indeed, there is a movement within the USMC to return to the M14 rifle and the 308 round, and, though I am in sympathy, that probably won't happen either. As I have reported, the American military is getting a new rifle caliber (6.8mm) and a new rifle to shoot it. We may end up with a new pistol also, maybe in 40S&W caliber.

In the interim, we all need to understand this: We don't carry pistols, because they're effective. We carry pistols, because they're convenient. When they are inconvenient (bulky, hard to operate), the point would appear to be lost. We carry pistols as a deterrent against unexpected threats. We carry rifles when we are facing expected threats. Rifles are not convenient to carry, nor are they concealable, if you add "ineffective" to that list, once again, the point will appear to be lost.

Pistols need to be convenient and concealable, marginally effective. Rifles are going, of necessity, to be inconvenient, so they need, at least, to be effective. It's a simple formula. One can only wonder why the American military bureaucracy, after all these years, still can't seem to get it.

29 Nov 2003

Holiday Report from LA:

"At 2200 hours, the combined levels of L-Tryptophan and Budweiser became elevated to the point where 'family and friends,' remembered, once more, that they really didn't like each other after all. My guys and I handled a rapid succession of out-of-control party and 'family fight' (now, there is a redundancy) calls that lasted well into Friday morning.

Our department's Glock transition schedule starts just after the first of the year for us patrol folks. So far, our Metro Division (includes K-9), Antiterrorism Division, and some other specialized/detective units have already changed over from the Beretta 92F. Curiously, every 'civilian consultant' our new chief has brought here from the East Coast is now carrying a Glock too. Though not sworn police officers, they carry a peculiar 'Police-Commission-issued CCW permit.' Funny, I didn't think we had a citizen CCW system here in LA!

Chief Bratton is popular with the troops! He has changed management philosophy at every level to be a whole lot more street-cop friendly than it has been. In light of the succession of dithering buffoons we're had in that post since Daryl Gates left, we, at long last, have a truly competent and qualified person as chief. We all hope he sticks around.

On the vest front, Second Chance is now replacing all of our 'Zylon' body armor with their standard, Kevlar vest. All vests are scheduled to be replaced within the next few months. As you noted, whether it actually works or not, Zylon is toast. We're all back to Kevlar."

21. Training Tips

8 Nov 2002

At a recent class in CA, all shooting was done one-handed. The curriculum was so well received that I'm doing a similar class at next year's IALEFI Conference in FL. Some painful conclusions:

One-handed shooting is not easy, and the expectation that, without practice you will be able to be accurate and sustain a reasonable rate of fire, is little more than wishful thinking

You must go back to basics and let your pistol do the work. There is a strong tendency to trigger bash in order to compensate for the reduction in stability and strength.

One-handed, weak hand shooting is extremely difficult, much more difficult than most shooters think, but it needs to be confronted and practiced too. When you have to do it for real, it will be too late to wish you had practiced!

Trigger-cocking pistols exacerbate the challenge.

You may have to use your non-dominant eye, an additional challenge to which few have ever given any thought.

28 Nov 2002

State of police training. From a friend who just attended a police training course. Attendees were all working patrol officers, ranging from rookie to veteran:

"The good news is that accuracy was acceptable in most cases, and most officers displayed a healthy, learning attitude. Here is the bad news:

100 % did not incorporate lateral movement into their pistol draw, stoppage reduction, reloading procedure, nor firing sequence. This matter improved by only 10%, even after movement was taught and

recommended.

40% consistently looked at their holsters during the reholstering process.

20% panned their support hand with their muzzle during the draw sequence and during reholstering.

20% were apparently comfortable holstering an empty weapon after completing each exercise, even though they knew they were on a hot range.

100% failed to look all the way behind them after firing and prior to reholstering.

75% failed to scan to any degree before reholstering.

40% were unable to reduce a stoppage quickly.

30% 'scooped' their pistols during the draw (sometimes called 'bowling')

75% of the students equipped with pistols with decocking levers were unable to decock rapidly using only their strong hand thumb. In many cases, these officers struggled and dithered every time they tried to decock."

Lesson: We're a lot better than we used to be, but we still have many training challenges that must be overcome. All trainers need to work diligently on the above issues.

10 Jan 2003

From a friend south of the border:

"We train law enforcement officers throughout Central America where the IMI Galil rifle (223) is what most police here are armed with. In fact, the Galil is the primary arm of police officers. Pistols are carried only by the brass.

For all practical purposes, rifles receive no maintenance, but, to their credit, they keep running. Unfortunately, officers fire only a few rounds during training and are then thrown to the wolves.

I almost never handle a Galil without losing skin! Lots of rough edges and sharp corners.

You may find it interesting that ten meters is the maximum range anyone practices here. The rifle is actually used as a big pistol. A conventional sight picture is rarely taught. Officers typically look over the rear sight and just use the front sight as a reference. A great deal of potential inherent to the rifle is never exercised."

Lesson: Small deeds done are better than great deeds planned. Trainees (military and police) will do exactly what is expected of them. When little is expected, little will be accomplished. Seldom are noble deeds done by men who are thought of, and think of themselves, as nothing more than cannon-fodder. Nobel deeds are accomplished only by noble men. Training is largely meaningless when it does not make the trainee aware of his mission, of his place in history, and of his own magnificence.

27 Feb 2003

From a friend in Tucson:

"Tonight's local news: Tucson police are trading in their H&K USPs (40S&W) for Glock 22s w/Surefire lights. Quantity is one thousand.

The newscast then showed a police instructor demonstrating the draw. He actually instructed his officers to contact the trigger and '...begin taking up the slack,' before gun is even at eye level. Apparently, they shoot every time they draw! There was no mistaking it. The audio quality was perfect."

Comment: It's not just ignorance. It's arrogance. Enlightened trainers stopped "prepping the trigger" thirty years ago. Difficult to believe this kind dangerous, outdated rubbish is still being taught, with a straight face yet.

From the cowardice that is terrified of the truth,

From the lethargy that is content with only part of the truth,

From the arrogance that thinks it knows all truth,

Oh God of Truth, deliver us!

11 Nov 2003

Timely advice from a friend and LEO trainer:

"We did the Tueller Drill. We started with students at low ready and had would-be assailants run towards the back of the range. Students were able to aim and fire at paper targets well before the runner covered twenty feet. However, once we had thus bolstered everyone's confidence, we repeated the process with students drawing from concealment. Seeing students struggle to draw pistols through down, fleece, and other warm clothing was an eye opener for everyone. My fellow instructor tried it with me playing the part of the attacker. I managed to run completely out of the range before he could draw his weapon!

As winter descends on us, it pays to think about how we are going to draw, in a timely manner, from under our winter clothes."

Lesson: Our most dangerous enemy is denial. We must be careful not to con others and never to con ourselves. Self-deception is a killer!

12 June 2003

More from the IALEFI Conference in Orlando, FL:

Dave Grossman has a knack for putting into words what so many of us have danced around for years but to which we've never attached descriptive labels. Dave and Gary Klugiewicz have both expressed the concern that too much of the training, ostensibly designed to prepare students for emergencies, is done in the abstract. They go on to say that, at some point, training must be conducted under stress, so that it will be useable to the student when he must apply it under stress. It's called "Stress Inoculation," and it is the term we've been looking for.

Training Tips

Psychomotor skills that are intended to be useful to a student in an emergency must be pushed from the frontal lobes into the midbrain. Thus, in training, stress needs to be manufactured, so that the student can be immersed in it as he trains. As I said, most of us do it now. We just didn't know how to describe it. Thanks to Gary and Dave!

Another of my colleagues brought up a second point that I wish I had thought of and articulated before now. In the middle of a gun battle, it is likely that the shooter will neither hear the sound of his own shots nor feel the recoil of his pistol. Tunnel vision and auditory exclusion have been documented for a long time now and are familiar to all of us. In fact, many are the instances where an officer has pressed the trigger on his empty revolver multiple times before realizing that it was not firing anymore and has, long since, needed to be reloaded.

We should, of course, be focused on our pistol's front sight, but we need to make it a practice of mentally noting, in our peripheral vision, the ejected case flying up and out. So long as we note the case departing the outline of the slide, we can be confident that the pistol just fired, even when we can neither hear the report nor feel the recoil. Again, it is something many of us are already doing. Now we know why it is important, just how important it is, and why our students need to make a practice of it.

Part 4: Regional Experience

22. Military

7 Mar 2003

War in the Mideast:

In the last century, this country has paid a terrible price for failing to finish wars. Most wars don't end in unconditional surrenders. Most wars fizzle-out inclusively (only to resume years or decades later), because both sides get tired of fighting and lose their resolve.

The price paid for unfinished wars is what we now see going on in North Korea and in Iraq. Unfinished wars have to be fought all over again. For sixty years US troops have had to be stationed in Korea, because that war there was never finished. Now, it is rearing its ugly head again and will continue until the war (which started in 1950) is finally completed. Then president Truman said of that war, "Our troops are not there to win anything." In so doing, he insured their failure and defeat. When we're not there to "win," we shouldn't be there at all!

Thirteen years ago, the Gulf War was also left unfinished. The president and congress lost their collective nerve and stopped just short of victory. Now, we're having to do it all over again.

Lesson: Don't adopt the tactics of losers. When the stated goal of any military conflict is merely the preservation of the status quo, there is no possibility of victory. For the sake of those doing the fighting, and for the sake of those who will have to fight the same fight all over again in the future when the issue is not resolved, the goal of all military action must be the unconditional vanquishment of the enemy. Anything less is a recipe for disaster.

13 Mar 2003

From a friend on active duty. Gun phobia continues on the front lines:

"I have now been here in Kuwait for two weeks and not once have any of us been allowed to insert a magazine (much less chamber a round) in any of our weapons! We have all been graciously issued a small amount of ammunition (for immediate self-defense, we're told) but we're prohibited from getting it anywhere near our weapons! Even senior officers are not allowed to have ammunition in their weapons. I use rubber bands to attach a magazine to my rifle. At least it is with the weapon and will be available quickly if needed. Of course, even that practice is frowned upon, but I do my best to make sure no one notices.

Sand blowing into empty magazine wells is a big issue here. We're told to keep the dust cover closed, but not to put a magazine in the weapon, even an empty one. It makes no sense, of course, but I, for one, am not willing to tape my magazine well shut."

Comment: If the training these guys have received is so great, why are they not allowed to be armed as front-line soldiers should be? Do we expect to be able to make an "appointment" for an emergency?

In spite of this universal "empty gun" policy, there have still been a number of injuries caused due to unintentional discharges in Army units, as well as with the Marines. So, with all these empty guns, accidents are still happening! One can only wonder if there would be any more accidents than there are already if all soldiers were trained, and expected, to carry and maintain loaded weapons all the time, and did so as a routine.

Not all dithering fools are confined to the UN!

27 Mar 2003

Progress of the War in Iraq:

When I was in Africa several weeks ago, the adult daughter of one of my friends asked me, "Do yo really think the USA can defeat Saddam Hussein's army?" Astonished that the question was even asked, I replied, "We'll squash him like a bug!"

Weeks later, it is obvious to everyone in the Hussein regime that they are being systematically squashed. They are facing the same problem the North Vietnamese did forty years ago, and they are putting their utmost effort into the same solution. Resigned to the fact that they cannot win militarily, they are actively recruiting allies in the American and European press in an effort to win the battle in the arena of public opinion. They are hoping they can make things so hot for President Bush and PM Blair that they can force a negotiated settlement, which will ultimately leave the Hussein regime in power, just as they did at the (premature) end the First Gulf War a decade ago.

To their everlasting shame, the American press is only too happy to oblige! American newspapers, from the LA Times to the NY Times, have become little more than the western extension of Al Jazzera. Same with in the BBC in the UK and in much of Europe. Same with American broadcast media, ABC CBS, NBC, NPR, and especially CNN (the Fox Network is the sole exception).

It is painfully obvious that, if Al Gore were president, all the press and media above would think this was the most wonderful war in history! They would all be for it. The press loved Clinton and his gang of sleaze. Without exception, they defended him in every scandal (a nearly daily event during the Clinton years) and in every military adventure. To them, Clinton could do no wrong. The loved Clinton precisely because he was sleazy, and he therefore reminded them of themselves.

President Bush is a decent, truthful, and honest person. His administration has not been plagued by scandal, not one. He attracts decent and dignified people to his staff, not sleazy trash like the former administration. He is not a trashy Marxist, and that is why Marxists in the press hate him so hysterically and so irrationally.

The press is not opposed to this war. They couldn't care less about the war, nor about our troops. They're just opposed to Bush. They'll predictably hate everything he does. They want their sleazy trash friends back in power, and they will even stoop so low as to join hands with the Saddam Hussein in order to make it happen. They are a disgrace to their profession.

Fortunately, the Bush Administration continues to stand tall, because those who would tear it down lack the stature to reach it.

2025 Update: These same Marxists in our American media hate Trump too, and for all the same reasons!

21 Apr 2003

From a friend with our forces in the Mideast:

"Got back in Kuwait on Easter Sunday. Over the last four weeks, we went all the way to Tikrit through Safwan, Nasiriyah, Numaniyah, Kut, and Baghdad.

Everywhere we went, we were greeted enthusiastically by local Iraqi people. In fact, the only significant resistance we encountered was from 'Jihadists' from Jordan, Chechnya, Syria, Palestine, Yemen, and Iran. I'm not sure what they were expecting, but they ran into a buzz saw! All are now in paradise with Allah, where they received, I'm sure, a warm welcome. It was surely warm when they left! Nearly all members of the 'vaunted Republican Guard' and Ba'ath Party fled like mice before we ever got close to them.

USMC Cobra helicopters performed superbly. They cleared the way all the way from Kuwait to Tikrit. Nothing stood up to them. They annihilated tanks, vehicles of all descriptions, artillery and mortar tubes, bunkers, troop concentrations, you name it. When we followed up, there was usually little remaining, save cinders!

I shot a gazelle in one of Saddam's private game preserves. One shot from my Beretta M9 pistol dropped him at a range of twenty meters. I used Federal EFMJ. No exit, and the animal went right down. He sure tasted good, particularly after thirty days of MREs.

I toured one of Saddam's Palaces. It is opulent beyond belief! All this while most Iraqis live in mud huts and barely eke-out a living.

The Army is now taking over the humanitarian phase or the Campaign.

Unfortunately, many Army troopers (who missed out on the real excitement) are now shooting at destroyed Iraqi tanks and artillery pieces, all the stuff that we Marines demolished weeks ago. It is unnecessary and scares the snot out of local civilians. I surely hope this current 'transition' goes better than was the case in Haiti and Somalia."

Comment: Makes me want to get back in uniform!

5 May 03

After-action report from a friend in The Theater:

"Most of the killing of Iraqis on the ground was done with 50 Cal and 7.62 Cal machineguns. Marine Corps employment of fire and maneuver warfare proved so lethal (and thus so intimidating) that the will of the enemy was nearly always broken before any significant ground engagement. F18 Hornets and AV-8 Harriers pounded them relentlessly, and then AH-1 Cobras annihilated what was left so completely that most engagements on the ground were simply mopping up, with the exception of several occasions where we were charged by literal human waves of Jihadists. However, machineguns and rifle fire quickly wiped them out. Charging our kind of firepower with foot infantry over open ground has no chance of success, no matter what the numbers. A lesson they learned the hard way!

There was some tough street fighting in Basra, Nasiriyah, and the outskirts of Baghdad, but once more, AH-1 Cobras significantly reduced our casualties. I talked with one Cobra pilot. He indicated that flechette rockets worked superbly on Jihadists in the open and on rooftops. Jihadists using building, rubble, and vehicles for cover were easily eliminated with 20mm cannon."

10 May 03

Small arms feedback from the Gulf War from a SpecOps friend in the area. This is confirmation from the First Gulf War:

"The M855, steel penetrator, 62grain 223 round, combined with the short,

M4 rifle has, once again, established a poor reputation. Most of our opponents were skinny and lightly clothed. None wore body armor. At close range, our rounds easily penetrated through and through, but cases of immediate incapacitation were rare, even when they were hit several times in rapid succession. Many complaints on this issue. The M855 does penetrate solid barriers, but no better than the Soviet 7.62X39.

At long ranges, the M855 round does not have enough power. At those ranges solid hits did not take people down quickly. Damage was disappointing. Again, there were many complaints.

Our soldiers and Marines did their job. They hit and hit consistently. The M855 ammunition failed and failed consistently. An upgrade is long overdue.

With this experience, we know all have a strong opinion that a better caliber is needed, before the next war! Right now, there is a movement underway within US Army Special Operations Command to develop a SOF Combat Rifle (our next individual rifle) for Special Forces. When asked, we all expressed the opinion that we need better terminal performance on people, at all ranges, right out to five hundred meters."

Comment: In Vietnam, our opponents were also skinny and lightly clothed, but a single hit from an M16 shooting 55gr hardball almost always took them down in short order. The 55gr hardball round we used then was limited to 150m in range and didn't penetrate well at any range, but it did a nice job otherwise.

Evidence suggests that the new, short M4 rifle, combined with the 62gr "penetrator" round fuses the worst of both worlds! Penetration is nothing special, and terminal performance is poor at any range. Let us pray that bureaucrats stop dithering and correct this obvious deficiency before the start of the next war!

18 Sept 2003

A faint and distant voice from the front:

"Our guys are getting good hits during CQB, with both M-4s and M9 pistols, but the bad guys are just not going down. They usually die eventually, but, in the short term, they stay on their feet, stay conscious, and continue to shoot at us. We have been, and continue to be, extremely unhappy with the poor terminal performance of both these calibers and issue ammunition, and we have tried to give voice to our unhappiness, particularly in view of the fact that this same problem has been well known since Somalia, and nothing has been done in the interim.

We do report these failures through the chain of command, but uncheerful information always gets filtered out and watered down. The end "report" describes us all as happy little campers. However, the bottom line is that the 9mm FMJ and the 5.56mm 62gr 'penetrator' are dismal failures as military ammunition, no matter what Pentagon atta-boys try to tell you."

Comment: We should not be surprised when internal "studies" reveal that all equipment issued to our soldiers and Marines works great and just couldn't be better. Such "reports" are indeed comical in view of the above. The only real purpose of these "reports" is to enhance the careers of the people who arrange to have them written (with all the "conclusions," of course, known in advance).

When I was in Vietnam thirty-five years ago, the 55gr hardball round that we used in our M-16s was, at least effective in taking the fight out of bad guys, so long as it didn't have to go too far, nor penetrate anything substantial. The new 62gr "penetrator" still doesn't penetrate anything, and the over-stabilized bullet is a failure in ending fights.

Of course, 9mm hardball is a joke, as we've all known for many years.

It was a foolish and inexcusable mistake to ever select 9mm pistol and 5.56mm rifle as military calibers in the first place. The Pentagon still doesn't want to face facts. Indeed, they don't even want to know the facts! While they continue to dither, brave men die.

24 Sept 2003

From a friend on active duty:

"Based on the Iraq experience, the DAMPL (Department of the Army Master Priority List) is out. Here are the highlights:

There are no more CAT 1 (really ready) or CAT 2 (sort of ready) units. Only units that are deployable immediately and those that are not. One of the reasons the 507th Maint Co got lost and ambushed in Iraq is because it got shoved in to the action without much of its equipment. It never should have been deployed at all.

On short notice, we are going to have to be able to deploy something other than just light infantry. Heavy backup is going to have to be (for the first time) instantly deployable too.

The division HQ, as we have known it, is history. In its place we need a task-oriented, modular command and control unit, instantly restructureable, instantly mobile.

An Army leader who balks, drags his feet, or complains about working in a joint/coalition environment will be fired. Political autonomy is history.

For once, we need to be honest about the capabilities of our helicopters and stop flying them over heavily defended positions without support and without any real chance of recovering downed aircraft and crews.

Unit cohesion is critical. We need to stop the practice of individual replacement and keep units together. When a unit is no longer combat effective, it needs to have its mission adjusted accordingly or be cycled to the rear.

We're getting out of the installation management business. Property management companies take better care of on-post housing and facilities than any housing office ever did."

Comment: To its credit, the Army is trying, under the prodding of Donald Rumsfield, to correct and upgrade, admitting (some) mistakes and trying to find real solutions. I'm not sure I agree with all of the foregoing, but progress is being made. It can't come too soon!

13 Nov 2003

The Jessica Lynch myth, from a friend in the business:

"As of today, I've seen six different accounts of Jessica Lynch's "story," including a syrupy, made-for-TV movie. I've seen this all before. The party in power is heralding her a hero. Disparaging remarks from the opposition and their benefactors in the media are called politically motivated propaganda.

As nearly as I can discern from various brainless, over-dramatized, spin-ridden accounts, her group really screwed up. If she herself couldn't get her M-16 to work, maybe she needs to take better care of it, or (heaven forbid) actually load it. She apparently spent most of the conflict curled in a fetal position, posing a threat to no one, and nearly everything else is undeterminable, because she passed out. Victim? Maybe. Hero? Hardly.

There is no doubt Miss Lynch is a fine person. However, I am weary of transparent lies being spoon-fed to me by both the government and the media. I am resentful. I prefer to know what really happened, minus the spin. If nobody knows, I prefer being told that nobody knows and probably never will.

If all they want is my political support, they can go screw themselves!"

Comment: None required.

23 Nov 2003

From a friend currently deployed in Afghanistan:

"I made it to Kuwait and Iraq before they sent me here. The differences in attitudes towards personal weapons among the three places is worth noting:

Guns & Warriors: DTI Quips – Volume 2

First, upon arrival in Kuwait, we weren't allowed to possess any rifle ammunition at all. Smart ones among us all carried (concealed) handguns, but the practice was, of course, frowned upon. Nonetheless, so long as pistols were not visible, no one seemed to care.

Suddenly, they graciously permitted us to keep a single rifle magazine on us, but, of course, we were told never to insert it into our rifles. That curious policy made things interesting when a local insurgent (whom the Kuwaiti government insists don't exist) decided to drive his car into the line of soldiers waiting to get into the PX. Fortunately, some of those soldiers saw it coming and loaded in time to pepper him, but not before he injured ten of us. The driver survived (chalk up another failure to the 62gr 'penetrator'), and was treated at the camp hospital until he was well enough to hand off Kuwaiti police. He is probably not enjoying his stay with them!

When I arrived in Iraq, we were issued our basic load and then we all locked and loaded, once we were past the berm that divides Kuwait and Iraq. However, back in a 'secure area,' we were required to unload our weapons, at least the ones they could see. We did get to keep our ammo. By this time, we veterans were all 'carrying concealed,' everywhere and constantly.

I then returned to Kuwait in order to go to Afghanistan. While there, I had the misfortune to pass through Camp Doha, where all the West Point pretty boys hang out. Saigon during Vietnam must have been something like this! We newly arriving GIs came in from Iraq dirty, smelly, and, to the horror of all the antiseptic and meticulously manicured folks there, bearing arms. They were aghast! The first thing we had to do was rid ourselves of those evil, dangerous things. So, we (once again) obediently divested ourselves of all our weapons (that they could see).

The command attitude here in Afghanistan is much improved. They want us to carry our rifles in transport mode (empty chamber, magazine inserted) in 'safe' areas, and we can carry all the ammunition we want. Indeed, I found the sight of a M249 with a hundred-round box of ammo,

Military

carried by a fellow soldier in line at the PX, to be comforting. Outside the wire, it's as it was in Iraq: full gear, locked and loaded (carry mode).

As in most wars, there are some bright spots here, but the welfare and safety of soldiers is the last thing anyone seems to consider. As you constantly remind us, 'we're on our own!'"

Comment: My friend is right. Vietnam was exactly like that! Soldiers were, and still are, trained only to operate weapons. They are never trained to live with them. Despite the best efforts of the "pretty boys" however, as we can see from the above, we are finally getting competent personal weapons training to some of our soldiers and Marines, making them into professional gunmen, not just "gun-operators"

26 Nov 2003

On personal weapon effectiveness from a friend in Iraq:

"Colt M4, 55gr FMJ 5.56, no problems point blank to 200 meters (our unit has gotten rid of the 62gr 'penetrator'). Most of our engagements involve multiple shots against each adversary. I'm not sure multiple shots are always needed, but it is so easy to just stay on target and keep shooting until he is out of the fight. Upper thoracic hits will get people down and out of the fight the fastest. Nothing new here. 5.56 55gr FMJ renders good penetration through vehicle windows, but my only experience is at under ten meters, and all shots were straight-on.

I bent the trigger guard on one of our favorite gunsmith's Lightweight Commanders. The trigger housing was bent and the underside of the frame was also slightly deformed. I shot this guy several times (45ACP Hardball, all we can get here), but he still would not let go of one of my people. I then grabbed him around his neck and beat him with the pistol. He finally succumbed to his wounds. I didn't notice the damage to my pistol until I cleaned it that evening. With a little 'home gunsmithing (vice, hammer, rebar),' it still works, and I'm still carrying it.

In another incident, one of my guys got hit (luckily, in the plate of the vest he was wearing) with a 9X19 pistol round at close range. He immediately returned fire with his M9 (issue 9X19 hardball) and hit the guy five times close to the body midline. All hits were above the waist: one in neck. The bad guy was still able to close the distance, grab my guy, and try to choke him. MP came up and pumped two 12ga rounds (00Bk) into the bad guy him at pointblank range. That finally ended the fight.

John, this is a wild place. Just as it was with you in Vietnam, we have contact every day. Every night the mortar rounds and small arms fire never stops. I saw a rifle round come through the tent and go through my sleeping bag and cot. I was sitting on a chair watching a DVD at the time. I saw another round come through our chow hall trailer, hit the salad bar, and then spin around on the floor. A trooper came up, grabbed the round, and said 'Lucky this thing didn't land in the salad; somebody could have broken a tooth!' Wartime humor, it seems, never changes."

Comment: 55gr hardball still works, at least at close range, and so long as you don't have to penetrate anything substantial. 9X19 hardball out of a pistol doesn't work well at any range.

All this we have known for years, but we still can't seem to get effective calibers, ammunition, and weapons into the hands of our soldiers. As I have noted, many have grown weary of waiting and have started to equip themselves.

2 Dec 2003

This is a portion of note sent by an active-duty friend to his junior officers. They will be deploying shortly:

"For those 'who don't know what they don't know,' I offer the following explanation: 55gr FMJ (Full Metal Jacket) is the old M193 ball ammunition that we used to issue before the new, 62gr FMJ (M855) was issued with M16A2 rifles. The M855, 62gr bullet has a steel core 'penetrator' (a hard dart imbedded in the middle of the bullet) and is boat tailed at the base to provide maximum stability in flight. The 1:7" twist

Military

(or, one complete revolution in seven inches of travel down the barrel) of the M16A2 barrel is severe, by any standard. The bullet revolves nearly three times before departing the barrel of an M16A2 (20" barrel), and two complete revolutions from an M4 (16" barrel).

This provides great stability for the bullet during flight, too much as it turns out, because the bullet uniformly fails to subsequently destabilize and tumble upon impact with living issue, as did the old, M193 bullet. It is basically a fast-moving ice pick. While it will make them bleed, it fails to generate a large, permanent wound cavity.

While the M855 bullet (with its imbedded penetrator) will penetrate some solid barriers that will not be penetrated by the M193 bullet, it consistently fails to penetrate most layered barriers (car doors, cinder block, etc) at any range.

Since, when we're deployed to Iraq, we probably won't be able to get sufficient quantities of the old M193 ammunition, and the M193 bullet renders poor accuracy (and sometimes self-destructs) from a 1:7" barrel anyway, we will be using the M855 round for the foreseeable future. We will therefore have to train our men to plan on hitting bad guys multiple times and not to expect much in the penetration department. That will be the case until we're issued the new caliber (or reissued M14s), which may be months (or years) away.

I can go into another diatribe about our 9X19mm FMJ pistol ammunition and its notoriously poor terminal performance also, and, yes, we're looking forward to improvements there too. In the interim, constantly remind your men and yourselves that the Marine is the weapon. The rifle or pistol are merely tools. The poorest performing bullet in the world, when it hits, will render infinitely better results than will 'wonder bullets' that miss! "

Comment: My heroic friend is trying to make the best of an unhappy situation. We were told that the new M855 bullet would solve both the range and penetration limitations of the M193. As was the case with the Patriot Missile in 1991, we were (knowingly) lied to!

The Patriot Missile didn't work back in 1991, but, generals and presidents alike spent too much time reading their own press releases! It was only when Patriot batteries failed utterly to protect Tel Aviv from Scud missiles that the lie began to unravel. We later learned, to our horror, that the success rate of the Patriot system in intercepting and destroying incoming Scud missiles was, in truth, close to zero!

By the same token, the hardened penetrating dart, imbedded within each M855 bullet, weighs only ten grains. There is no way a ten-grain projectile is going to penetrate anything! The truth is, neither range nor penetration limitations have been adequately addressed by M855 ammunition. It is woefully inadequate on both counts and vastly inferior to the M193 round it replaced.

I'm not sure I'm the world's expert on ballistics, but I have been an infantry officer involved in significant and continuous fighting, and I know that rifles carried by infantrymen need ammunition that will shoot through things! Even in the USA, there still no species of political alchemy that will render "golden" performance from "leaden" ballistics!

8 Dec 2003

Readiness comments from another friend and student on active duty:

"I spent last week training and evaluating DOD Security Police at a major military installation here in CONUS. What a contrast! Attitudes of all the participants were excellent. They were all eager to learn and willing to listen. Unfortunately, gun-handling skills, marksmanship, and tactical skills ranged from marginal to abysmal. They have the hearts of warriors, but not the skills.

These kids are required to carry their M9 pistols with an empty chamber and the manual safety/decocking lever down. Many don't even have a magazine inserted. They are issued Level III retention holsters and are required to carry only hardball ammunition. They are given no training in drawing the pistol from the holster, reloading, or stoppage reduction. The only range time they get is for 'qualification,' which is, of course, a joke.

I observed that most of them are so unfamiliar with their equipment that, upon drawing the pistol from the holster (those that can actually figure the holster out), they fail to chamber a round, fail to take the safety off, and routinely point the muzzle at their own hand. Grip, stance, movement, verbal commands, and administrative skills have obviously never been practiced or even discussed.

These kids have no access to rifles, nor machineguns. The pistol is all they have. Senior NCOs and officers here readily recognize the glaring deficiencies, but none of them have the necessary skills either, and they don't know where to go to get them. Bureaucratic inertia, of course, encourages them to do nothing. Few are willing to 'rock the boat.'

Lesson: The blind are trying to lead the blind. Competent training in critical weapons skills is available, but the blind are too proud to ask.

23. Philippines

4 Nov 2002

Rifle information from a friend in the Philippines:

"Over here, the M16 series continues to enjoy respect. Government units equipped with the M16s still have the old variant with the slow twist rate. All use the M193 ball round. Contact distances remain well under 200 meters, with the vast majority of encounters occurring within 100 meters. Given all this, the (old) M16 still shines around here.

Some 'elite' units (devoted to protecting politicians) are now fielding a new, 'mini' version of the M4. Iron sights have been replaced by red dots, with no backup sights. Barrels have been shortened to six inches. Stocks are maladroit, retractable wire contraptions. Rifling is 1:7. Considering the length of the chamber, one has no more than four inches of rifling to stabilize the round. Consistent zeros are impossible to establish beyond fifty meters. The maker is blaming it on poor marksmanship. Next project? A four-inch version of the same gun, ostensibly for those really close encounters! Heaven forbid these folks run into well trained riflemen, using real rifles.

Meanwhile, our Rangers are content with their standard M16A1s. Those equipped with M14s are even more delighted. Not surprisingly, these are the guys who've seen the most action in the most places, urban, field, mountain."

Lesson: Heaven save us from "elite" units! The 223 round in a standard-length rifle is effective to 150 meters. Beyond that, its effectiveness diminished exponentially. Its penetration capability (at any range) is extremely limited.

Shorten the barrel to under sixteen inches, and a rifle which is already effective only at relatively short range, becomes even less effective.

Guns & Warriors: DTI Quips – Volume 2

6 Nov 2002

Shooting Incident in the Philippines:

"It was early in the morning. I was returning from a golf outing in a taxi. The driver stopped by the side of the road to relieve himself. I decided to do the same.

Three men approached us from across the road. Sensing trouble, I hurriedly finished and started moving away from the taxi driver. One of the three asked us if we lived in the area. Not waiting for a reply, another announced a holdup. Two pulled pistols. The third brandished a large knife.

Knowing we would both probably be killed no matter what we did (the typical armed robbery MO around here), I moved and simultaneously drew my own pistol (1911 w/hardball). I ended up on the opposite side of the car as the robbers.

We traded shots through the car windows, spraying glass fragments everywhere. I fired widely at first, hitting nothing. When I finally settled down, I hit one robber twice, both times in the chest. He was dead at the scene. I, in turn, was hit twice (38Spl RN Lead), once in the leg, and once in the jaw. The leg wound was through-and through. The jaw hit broke my lower mandible but did not penetrate my neck or head. I felt no pain from either would. In fact, I only became aware of the leg would after the robbers fled.

I knew I was hit, but I also knew the fight was not over. I applied pressure to the facial would with my left hand and continued shooting with my right. When I reloaded (which required both hands), blood spurted out the entry wound.

I carried one magazine in the pistol and two spares on my belt. I was into my last magazine when the taxi driver, also wounded, got inside the cab, retrieved an AK-47, and started firing it full auto at the robbers. He had no knowledge of how to use the weapon, and most of his rounds were

Philippines

launched into the air, doing no damage, but it enough to scare the two robbers who were still alive, and they fled. Both the taxi driver and I then headed to town. I passed out just as we arrived at the hospital.

Oral surgeons reconstructed my jaw. It's been several months now, and I am able to eat normally. I'm lucky to be alive.

I carried my 1911 with the hammer down on an empty chamber, because I was told it was 'safe' that way. Never again! I now carry cocked and locked. I'm suddenly serious about living!"

Lessons:

Most robbery suspects flee like mice when the first shot is fired at them. However, sometimes criminals are willing the shoot it out, as happened here. Some are cowards. Some are crazy. Some are just mean. Either way, we must be prepared to best them in a genuine fight to the finish.

This fight occurred across the width of a Camry, but it was dark and only portions of bad guy were exposed and only for short times. In addition, my student had to engage most of them using only one hand. We don't practice one-handed shooting nearly enough!

Most gunshot wounds are not fatal. In fact, if you are going to die, you'll be unconscious almost immediately. So, when you're still conscious after being hit, it means that your wounds are probably not life threatening! We have to refocus immediately on the task at hand. When we allow ourselves to dither, the problem will likely solve us.

Keep emotions in check. While my student's anger kept him in the fight, it also caused him to forego the basics of marksmanship. He panicked, forgot his sights, and started to bash the trigger wildly. When he finally settled down and used his sights, he hit the robber where he needed to be hit. The sooner you settle down, the faster you'll end the fight!

When you carry a pistol for defensive purposes, carry it with a full magazine and round in the chamber! When you carry a pistol with an empty chamber, you're kidding yourself. Get serious or go back to eating grass.

12 Nov 2002

IPSC Shotguns in the Philippines:

"There was a strong presence of hyper extended pumps, with thirty-inch barrels and extended magazine tubes that held as many as thirteen rounds. These shotguns are a full foot longer than a standard M16. Not surprisingly, those with these monstrosities had a difficult time maneuvering.

I used a standard Mossberg 590 pump that held five rounds in the magazine. At every stage, I was offered the use of a hyper-extended shotgun, which I refused. 'Don't you want to get a high score?' they all said, as if that was all life had to offer.

In the high round-counts stages, guys with the long magazine tubes still had to reload, and, when they did, it became painfully obvious that they had not practiced. They dithered and fumbled, dropping rounds like rain.

Amidst all this, I was ordered to render my shotgun 'competition legal' by removing the sidesaddle and the sling. Seems the sidesaddles and slings aren't allowed as they project a 'military appearance.'

I don't really care if these folks want to deceive themselves, but it concerns me when they presume to train others in the art of defensive shooting, of which they obviously know nothing and in which they obviously have no interest."

Lesson: Competition shooting is the same over here. When all you want to do is just be "one of the boys," stick with the boys.

11 Dec 2002

Comments on military rifles from a friend in the Philippines:

"John, two rifles seem to stand out in your posts: the DSA FAL (308), and

Philippines

the Robinson M96 (223). I don't recall your using an AR much, at least in your posts. In fact, when we met in California several years ago, you were traveling with an M1 Carbine and a Remington 12ga 11-87P.

The reason ARs continue to be popular here is the abundance of spare parts, particularly magazines. However, I've seen all the AR issues you've mentioned materialize over here, in spades. I'd gladly change over to an RA-96 if we could get them.

Other rifles we see here:

The Galil shoots well, but, even in its abbreviated versions, it is quite heavy. Magazines are hard to find.

The SIG 550 is slim and smooth, but, again, magazines are expensive and hard to find.

The AUG works well. I'm skeptical of optics, but I'll admit that the one on the Steyr is easy to use. Trigger pull reminds one more of a pistol than a rifle. Once more, magazines are hard to find."

Comment: Back here in CONUS, the Galil and the Steyr AUG have been banned from import since 1994 and are today essentially unavailable. The SIG 550 is imported but can be sold only to law enforcement agencies.

Military rifles manufactured in the USA (and thus available for purchase by citizens in most localities) include the Robinson Arms RA-96, the DS Arms FAL, Springfield Armory's M1A (M14), and ARs produced by Colt, Bushmaster, DPMS, et al.

That is what is currently available new. Garands, M1-Carbines, and a host of others are still available in the used category.

Any military rifle that works well and can be maintained is an eligible candidate. We all need to make up our minds and get what we need.

7 Jan 2003

From a friend in the Philippines:

"I chanced upon a copy of Shooting to Live by Sykes and Fairbarn. Of course, it is dated, and the Commonwealth English, as well as windy writing style, must be overcome.

However, what struck me was that the nature of lethal conflict has not changed between when the work was written and the present. The authors' highlighting the importance of low-light skills and automatic stoppage reduction, as well as emphasizing that real criminals move and shoot back is as relevant today as is was then.

This stands in stark contrast to commercial gun magazines (and today's military philosophy) that dwell endlessly on hardware rather than software components of personal survival.

The philosophy that the book espouses is timeless. As you emphasized in your last quip, fearful men, particularly those who are afraid of guns, can never be victorious no matter how much 'training' they've received."

31 Jan 2003

Update from the Philippines:

"Our National Police just celebrated its 12th anniversary. In attendance was our president who, of course, congratulated the police brass (odd, because violent crime here is higher now than it ever was before their existence). In the very next breath, she ordered the chief to suspend, indefinitely, the issuance of 'Permits to Carry Firearms Outside Residences,' our equivalent of your CCW permits.

We were told this was being done in response to dramatic increases in violent crime (none of which is caused by legitimate permit holders). The logic is that, if only police are allowed guns, it then becomes easier to apprehend criminals, since criminals will be the only other category of citizens who have guns (did I get that right?). That leaves only three kinds of citizen here: armed police, armed criminals, and defenseless peons.

We peons shall now be limited to transport permits only, meaning guns have to be locked in trunks, unloaded, with ammunition in a separate case. To enforce the ban, we are told police checkpoints shall be set up nationwide.

Enough is enough! Already, the responsible, gun-owning public is organizing. On Monday, gun owners will troop to the Department of the Interior and Local Government in protest. Protesters will be wearing holsters with eggplants in them. History has taught us that no people, not even a nation of cowards (like this one), can be oppressed indefinitely. The people's patience with these capricious, self-righteous despots grows short. I only hope this does not lead to open rebellion, as it has in the past."

Lesson: To politicians (there and here), individual rights do not exist. There are only 'privileges,' which they may, at their whim, grant or withdraw. One only has to listen to a typical political speech to gain that distinct impression. A sound mind can be neither bought nor borrowed, yet unsound ones are being purchased every day. The world is going backwards.

Pluralistic democracy was once touted at a permanent cure for the arbitrary power of tyrants and dictators. After two hundred years of experience, we now see (all too plainly) that tyranny can exist within a democracy too, a phenomenon once naively thought to be impossible. The dignity and wellbeing of individual citizens is always the last thing considered by government officials, even elected ones.

17 Oct 2003

Sage readiness comments from overseas friends, one in the Philippines and one in SA:

From SA:

"As you know, we live on a farm and must be able to keep multiple attackers occupied, perhaps for as long as a day, before any kind of help is likely to arrive. I would love to carry a pistol chambered for 40S&W or 357SIG, but we're pretty much stuck with 9mm over here.

We now fit the red Glock firing pin Springs (twenty-eight Newtons, not quite sure of the pound equivalent) and maritime/amphibious firing pin spring cups to all our Glocks. We found these Glock-approved modifications significantly improve detonation of the hard primers that are so common here. Combine this with a NY1 Trigger, and the Glock pistol is unbeatably reliable. I carry a G17, loaded with WW147gr Black Talons (Cor-Bon when I can find them). Gwen carries a G19. At the house, we have several G18 (33rnd) magazines lying about. With accurate shooting, we should be able to destroy a dozen or so and hold off the rest. I'm acutely hoping I don't get the chance to test my theory!"

From the Philippines:

"The legal climate is strict over here, and we are only allowed to own one rifle. The predictable tendency is to find one that fits all tasks. Naive people are still insistent on having a single rifle that is simultaneously suitable for both CQB and long-range sniping. As you noted, such a creature does not exist. As in the States, most of the people here insisting on owning a rifle that will 'drive tacks at six hundred meters,' lack the personal competence to hit any kind of target at one hundred! If your rifle will consistently group into a twenty-five-centimeter (ten-inch) circle (under field conditions) at two hundred meters, it is sufficient for any challenge you are likely to encounter here.

In addition, our experience is that the first guy taken out by the bad guys is always the one with the shiny, gimmick-laden gear. Guys with 'high-speed gear' are perceived to be the more affluent and thus the 'alpha members' of the pack. Get rid of them first, and the entire group will fall into disordered chaos, or so goes the theory.

Civilian gun owners have had to defend their homes and villages from raiding insurgents. Sieges are typically fourteen hours, or longer, before government forces arrive. Some in remote areas have lasted for days, even weeks. Neighborhood skyline is typically composed of two-story houses and shops, surrounded by open fields. Most engagement distances are within two hundred meters, and fighting usually carries over into the night. Most shots are on moving targets.

Guns that see the most action include the ubiquitous M16, but also much 'obsolete' weaponry, including 30-06 Garands, 308 M14s, and M1 Carbines. Carbines are especially popular among women and teenagers. In these town raids, the carbine proved its worth more than once! Bad guys are particularly scared of the Garand, because it shoots through nearly anything, and it is effective at very long range. More than one surprised insurgent has been killed by a 30-06 bullet, when he thought he was far enough distant to be safe!

The foregoing stands in stark contrast to the manicured lawns and perfect target presentations typical of static rifle ranges, where increased difficulty is always created by increased distances and/or smaller targets (that still remain distinct in color against the greenery).

As you say, it's not a 'game' over here. Insurgents may not be competent riflemen, but they mean to take over this country, and, if you get in their way, plan on a desperate and prolonged fight!"

Comments: My esteemed colleague, Louis Awerbuck (now deceased), made a good point when he suggested that we complicate training challenges, not by increasing target distance, but by reducing contrast and adding movement. At any such suggestion, of course, all the "target-shooting kiddies" shriek in horror!

While, in this country, we make casual and quaint games out of critical shooting skills, good and decent people in other parts of the world are literally fighting for their lives, as we can see. We may face the same situation here, and maybe a lot sooner than any of us think!

20 Dec 2003

Comments on current history from a friend in the Philippines:

"Recent history has proven once again that, despite the advent of cutting-edge, siege weaponry, like bombers, missiles, and unmanned drones, rigorous ground campaigns are still nonnegotiable requirements. Rough men still have to physically assault physical objectives, and these men still have to be skilled, armed, and personally committed if they are to seize, maintain, and hold this ground. The desperate, continuing armed conflict in 'Postwar' Iraq should shock this 'revelation' into our veins."

Comment: The veins of those who naively think there is a technological substitute for personal tenacity and icy determination, engendered by personal patriotism, should be particularly shocked.

"With a ghastly casualness, we remove the organ and then demand the function. We make men without integrity, and expect of them virtue and enterprise. We laugh at honor, and are then shocked to find traitors in our midst. We castrate, and bid the geldings be fruitful."

CS Lewis

24. South Africa

25 Nov 2002

Latest from South Africa:

"As the festive season approaches, we are experiencing our seasonal increase in cash-in-transit heists. This year, a new tactic as emerged. I'm surprised the robbers haven't thought of this before:

The armored van is forced off the road in a remote spot. Most of these vans are top heavy and roll easily once they leave the road and hit the dirt shoulder. AK47s are then used to riddle the van's soft underbelly. The sides are armored, but not the underside. The guards inside are usually all killed. The robbers then can peel off the doors at their leisure."

Lesson: Never stop moving. Don't let anyone force you off the road.

14 Dec 2002

From a friend in South Africa on their new gun laws, going into effect sometime after the first of the year:

"This legislation was promulgated in order to vastly reduce the number of privately owned firearms. All civilian firearm owners will be restricted to owning only two guns. Even then, licenses will only be issued on presentation of 'adequate justification,' so government bureaucrats can still arbitrarily veto anyone they don't like, such as their political opponents or anyone who dares to speak out against them. Anyone with 'extra' firearms must turn them over to authorities for destruction or 'other uses,' without compensation. When a gun breaks, there is no provision for a person to borrow one while his is being fixed.

A 'reason' or 'purpose' will attach to each licensed gun. When a person owns a rifle for 'hunting,' but is forced by circumstance to use it to defend his home or himself, he will be liable to prosecution, because the gun was not used for its stated 'purpose.'

In addition, severe limitations will be placed on how much ammunition a person may possess, two hundred rounds per licensed firearm.

Firearms licensed for 'defensive purposes,' may still be carried concealed, however vast areas of the country are now being declared 'Gun Free Zones' (read that: 'Criminal Empowerment Zones'). Virtually every building will be a 'Gun Free Zone,' making the routine carrying of a defensive firearm virtually impossible.

Everything is directed towards legal gun owners, who aren't the problem and never were. As you can see, most provisions are unenforceable. Criminals aren't even mentioned.

We all know, and the government has admitted, that none of the foregoing will reduce crime, just as it has failed to reduce crime in Australia or the UK. What they care about, and the only thing they care about, is control, and when innocent citizens are murdered wholesale as a result, our politicians (much like yours), couldn't possibly care less."

Lesson: We must consistently distrust promulgators of such innocent-sounding schemes as "Gun Free Zones" and "reasonable" limitations placed on guns and ammunition. You can see where it is all leading.

We must all recognize that politicians positively love high crime rates. After all, the more crime you have, the more "government" you need! When crime is rampant, people are scared. Scared people vote for Democrats/Marxists.

19 Dec 2002

Incident avoided in SA. This is from one of my instructors there. He is also a local police officer:

"Last week I was the selected victim of a group of three would-be robbers. My wife and I drove to a local market to buy groceries. It was a particularly hot day, and I was wearing shorts and a loose top but had my CZ75, loaded with 115g Cor-Bon, in an IWB holster similar to what you wear. My wife was armed also. Standard procedure for us.

On arriving I noted three men standing next to the vehicle parked a few bays away. They seemed interested in something else, so my wife and I exited our vehicle and went into the supermarket to do our shopping. It is not uncommon to see these people in shopping areas. You would call them 'street people.' Over here, they are everywhere, and they can be extremely dangerous.

Upon exiting the market, I scanned and immediately picked up one of the men. He was loitering about looking shoppers over, including my wife and me. Suspecting that an assault awaited us in the parking lot, I directed my wife to go back into the supermarket, exit via the other side of the mall, and wait for me there. I then exited alone and entered a secondhand store next to the supermarket in an effort to see where the other two suspects had gone.

While in the secondhand store I inadvertently noted a fine selection of axes. Inspiration! I purchased a large one and left the shop with it in my hand. Immediately the one suspect I had spotted started to follow me, but, upon seeing the axe, he nervously sped up past me. I then saw that his two friends were waiting for him, crouched behind a car. They spoke and then all looked at me. All three then stood up and gingerly jogged away, never looking back. I picked up my wife on the other side of the mall and we were on our way.

Another aggravated assault (and probably a fatal shooting) averted. Here, it is something we have to deal with all the time."

Lessons: (1) There is no substitute for alertness. The most dangerous threat is the one you don't see!

(2) Have a plan! The ability to confront an unpleasant situation squarely, size it up, and make a plan on the spot is far better than stumbling into something and then having to make it up as you go along. Plans often have to be tweaked and sometimes even abandoned for another, but always having a blueprint in front of you will show you to the next step and

prevent panic. At the moment of truth, the best thing to do is the right thing. The second best is the wrong thing. The worst this to do is nothing!

(3) Go armed! Have the tools you need always at hand. Your next fight will be a "come as you are" affair!

30 Dec 2002

Violent Crime in SA:

"As usual during this time of year, there has been a dramatic increase in armed robberies here. The MO of robbery/burglary suspects includes brutally murdering victims, family members (of all ages), and all others in the house. This is, of course, an attempt to prevent identification later, but, in addition, violent criminals here know our government is really on their side. Our government likes people to be scared, and criminals are only too happy to provide that valuable service for our politicians. Violent criminals know that there is little chance of being caught and even less that they will be punished in any way that will disrupt their lifestyle.

When confronted by violent criminals, I have advised all of my (and your) friends to immediately and enthusiastically counterattack, at once applying maximum deadly force to all suspects, no matter the odds. As is the case in your country, our bad guys here do an amazing disappearing act when confronted by an armed citizen who gives every impression that he knows what he is doing with a gun."

Lesson: Weakness perceived is weakness exploited! "Prey behavior" on the part of victims always elicits "predator behavior" on the part of criminals. When you're legitimately ready, able, prepared, and willing to fight at all times, you'll probably never have to. As my first chief of police once said, "A little respect goes a long way."

22 Jan 2003

Present Situation in Capetown, SA:

"One of our city police officers was murdered Sunday. The officer was in uniform and at his residence waiting to be picked up by a colleague. According to witnesses, the officer answered a knock at the front door of his house, expecting to see his friend. Upon opening the door, the officer was confronted by two armed-robbery suspects. To his credit, the officer immediately attempted to disarm one of them. The thug's pistol (type/caliber unknown) fell to the ground. The thug's accomplice then stepped in, scooped of the pistol, and fired two rounds at the officer, hitting him in the upper leg, just below his ballistic vest (at an upward angle). The officer collapsed as the assailants fled the scene, firing a number a number of rounds back at the house as they ran. The officer bled to death before the ambulance arrived. Our city police officers are required to hand-in their pistols at the end of each shift. This murdered officer was unarmed at time of the assault. No arrests have been made, nor are any likely.

The next morning, four hundred city police officers staged a protest to draw attention to the fact that officers do not even have locker space at the precinct stations to change from/into civilian dress. This leaves them in uniform and unarmed to and from work. With many officers having to make use of public transport, they are constantly at risk. Our chief of police has been asked to resign, but no policy change is likely in any event.

Local politicians shed the routine crocodile tears and expressed their 'deep concern,' putting the blame for such horrible crimes on everything and everyone, except themselves. Some things never change!"

Lesson: This is what liberals want for American police! The obvious lunacy of unarmed officers running around in uniform is lost on liberal (Marxist) politicians. To them, police officers, like soldiers, are just so much cannon-fodder. For all their professed "caring." they couldn't possibly care less.

I tell all my soldier and police students to worry less about "rules," and more about living long enough to collect their retirement!"

2025 Update: In the intervening twenty-two years, Democrats haven't changed their tune, not one bit!

24 Feb 2003

Didn't take long! On of my LEO students from last week's class in Capetown, SA called this morning:

"Yesterday, just a week after attending your class, I was involved in a shooting here in Capetown. I confronted two escapees from our penitentiary system. As I learned in class, I assumed the interview stance and started moving laterally, as I issued verbal commands. The two suspects tried to get me between them, but I continually stacked them.

One picked up a brick and threw it at me. I moved, and the brick missed. I fired several shots, zippering him up as I had been taught (9mm hardball). I used my sights. All shots fired struck the suspect. None missed. He went down. His partner immediately surrendered and begged me not to shoot him.

One suspect DRT. One back in custody. No one else hurt.

If this had happened two weeks ago, I would have stood in one place, and I would have been carrying a pistol with an empty chamber."

Comment: Good show! This officer came to us on his own volition and on his own dime. At the moment of truth, he was ready and confident.

Victory!

25 Feb 2003

Hunting in South Africa 2003

Everything in Africa either bites, scratches, punctures, stings, or charges, and nothing in Africa dies of old age! This year, it was beastly hot in February, and comfort was surely not on the menu. This was the year of the "long shot," but, at the end of the day, it was my most satisfying African hunting expedition to date.

On the morning of 15 Feb 2003, I was in a massive game preserve north of Capetown with my friend and Professional Hunter, Joe DaSilva. We saw

South Africa

great herds of springbok, red hartebeest, oryx, bontebok, and blue wildebeest, as well as baboons. It's quite a place, and I was indeed fortunate to be able to take advantage of Joe's connections.

This day were looking for eland. Eland are among Africa's largest antelope. Light brown in color, their skin hangs on them loosely, like a wet towel.

Eland were elusive that day, but we finally located the herd for which we were looking. On foot, we tried to get close enough for a shot. I was using a borrowed Winchester bolt gun in 300H&H equipped with a 3/9X scope. Most hunting rifles in Africa have two-pound triggers, much lighter than I am used to, as I had to continuously remind myself. Failing to remember that would cost me dearly two days later. We slithered and crawled to within 150m of the bull we wanted. Brush was hip high, so we couldn't stand up without being seen. Wind was in our favor.

I acquired a sitting position, but I was still breathing hard. I was also hot, sweaty, stuck with thorns, harassed by bugs, and profoundly uncomfortable. Tormenting decisions are always confronting the big game hunter. Should I try to get closer? Is there something nearby that I can use for a solid brace? The herd is going to bolt before long, so I have to ether decide to take the shot or let it go. The fact is, you are never going to be completely satisfied with your situation, but circumstances will always force a decision. Any big game hunter who claims to have never muffed a shot is either lying or hasn't hunted much!

I decided that what I had then and there was as good as it was likely to get. It was getting late, and I knew I was running out of time. So, I settled down and did my best to hold of the point of the shoulder. The eland was standing in profile. My shot broke, and I heard it hit.

I immediately bolted in another round, as is my habit, and tried to get in another shot as the bull started to run. However, he immediately mixed in with others in the herd, so a follow-up shot was not possible. When my shot broke, it was low, but I thought it was still good. I was wrong, and I knew it as soon as the animal started running.

We looked in vain for the downed animal. He had, in fact, rejoined the herd, and we then had an arduous task before us. It took two hours for us to get into a position where I could take another shot at the same animal. The bull had been hit in the front leg. The bullet had splintered the huge humerus but had not broken it. This time, I had a firm brace. Range was 175m, and the animal was again in profile and standing in the middle of the herd. I had to wait until I had a clear shot. When my shot broke, I again heard the bullet hit. This time, I was confident the crosshairs had been on the point of the shoulder, but, of course, I bolted in another round and got ready to hit him again. Joe put his hand on my shoulder and said gently, "It's okay John. He's going down. Great shot!" Joe was right. A second shot was not necessary. The bull weighed in at 1,200 lbs, the biggest animal I've ever shot.

The next day, I was on my way to the Karoo Desert in the central part of South Africa with Ian, another friend and Professional Hunter. We spent the evening at a remote hunting camp, and I managed to get stung in the face by two extremely aggressive African bees. One nailed me just below the left eye, and I had a nice black eye the next morning. Nothing is this business is predictable!

We were up early the next morning on the trail of a herd of blesbok. The Karoo is rocky, devoid of vegetation (except in stream beds) and hot. Successful stalking requires one to take every conceivable advantage of what hills and rocky outcroppings there are. It took two hours of careful stalking to get within 300m of the herd. Ian identified the lead bull. This day, I was using a borrowed bolt gun in 308 with a fixed, 6X scope.

I had a firm brace this time, but heavy breathing, discomfort, and all the other ills to which flesh is heir made their presence known. I held on the animal for an eternity, waiting for a clear shot. Once again, I was not happy with the situation, but it was painfully obvious that we were not going to be able to get any closer, and the herd was moving and would be out of range before long. "Are you going to be able to make the shot?" said Ian. "I don't like it," I replied, "but there it is, and it doesn't look as if it's going to get any better."

I held a carefully as I could, again on the point of the shoulder. When the shot broke, it felt good, and, again, I could hear it hit. The ram staggered and fell within a few seconds. I was ready to shoot again, but it was not necessary. The 150gr Hornady bullet had gone through and through. I breathed a sigh of relief. Two down!

My next opportunity came a half hour later. Steenbok are small antelope, like klipspringer and dyker. I've seen them before but never had a shot at one. We jumped a nice one, and he ran for 100m and than stopped in profile. My crosshairs were high and coming down. I touched the trigger too soon, and the shot broke with the crosshairs still above the animal. He scampered away unscathed. I cursed myself for making the very mistake about which I always caution my students. I had my finger in contact with the trigger too soon.

No time to lament over lost opportunities. We were immediately off to another area, this time looking for the elusive black wildebeest. Two years ago, I shot a nice blue wildebeest, but the black variety is more difficult and much more dangerous. Both blue and black typically display the characteristic clown-like behavior that is endemic to their species, but the black variety has a reputation for charging unwary hunters, much as does cape buffalo.

Stalking was again a challenge. The shot was again 300m, but this time the animal was facing me. I asked Ian if we could get closer. He replied that we didn't want to get any closer! Ian said to me, "John, you can make this shot. Hold right on the nose. The bullet will drop a few inches and hit him in the chest." That is exactly the way it happened! I heard the shot hit, and the animal started running (away from us, thank heaven!), but I could tell he had been struck solidly. We found him 50m from where he had been hit. Fine specimen!

My last shot that day was on a springbok. He was running, and I was waiting for him to stop. He stopped at 300m, but not long enough. When he stopped the second time, he was standing in profile at 400m. Once again, Ian said calmly, "John, you can make this shot. Hold a foot over the shoulder. The bullet will drop right into him." And, so it did! After two seconds or so, I could hear that the bullet had hit. The ram flipped over and never took another breath. What a way to end the day!

So, I ended up with four wonderful animals. My performance was far from perfect, but I gained valuable experience (and a little humility). Hunting is always a dicey mixture of emotions and outcomes. Next time, I will be hunting kudo and zebra, and (if I can afford it) cape buffalo.

1 Apr 2003

From one of my instructors in South Africa:

"I've just finished presenting a five-day Tactical Firearms Instructor's Course here. Students fired 650 rifle, 650 handgun, 300 buckshot, and 10 slugs.

Mini-14 stopped working after 250 rounds. It lost its aftermarket flash hider (blew off the front). These rifles always gives us problems. Not my favorite.

Walther PPK 7.65 stopped working after 100 rounds. Terrible gun.

Rottweil semiauto shotgun stopped working after thirty rounds. Absolute junk!

Glocks all worked 100%

'R' rifles (South African version of the Kalashnikov) worked 100%

CBC shotguns (Remington 870 copy) worked 100%. Sights are poor.

Leather holsters all turn to goo here sooner or later. Most of us here have now gone to KYDEX, to which you introduced us!

Coaxial-mounted flashlights on rifles and shotguns are a Godsend. Surefires work well.

South Africa

One more topic: Bless your guys in Iraq. The good people in South Africa (there are a lot of us!) are backing you 100% !"

24 Apr 2003

From a friend in South Africa:

"In 2002, 9,498 people were murdered in South Africa! Twenty times that number injured in felonious attacks. Our murder rate is roughly one hundred times what it is in the USA! Our government is doing nothing to bring this figure down. In fact, all they can talk about is further restricting private ownership of firearms. This Sunday past, we had fifty armed robberies. Most suspects were armed with AK's and R's. Most all are stolen from the government or illegally imported (no private ownership involved).

Our politicians have come to the realization that 'The more crime you have, the more government you need!' The violent criminal element has, in effect, become a branch of government. Their job is to terrorize the public, so that the public will want a bigger and more intrusive government. Like politicians everywhere, none really care what happens to the nation, so long as they stay in power."

13 July 2003

Victims as Heroes?

Statistically, the most dangerous occupation in the world today is that of being a farmer in South Africa. Three of every one thousand will be murdered this year. Most victims will first surrender meekly to their home's invaders. They will then be tied up and tortured unmercifully before ultimately being shot or slashed to death. It is "ethnic cleansing," of course, but it is currently being ignored by the rest of the world. Crocodile tears are (at least in public) shed by officials, but nothing is being done to address the "problem." The entire Jewish population of Nazi Germany in the 1930s went through something similar. No one cared about them either.

In our cowardly unwillingness to face such historical events squarely, we, as a civilization, have waxed delusional. Instead of reminding every citizen that he or she must take reasonable precautions to avoid criminal victimization, we have, instead, elevated the moral standing of victims to that of sainthood!

We are, for example, assured by grass-eating bureaucrats, even police chiefs, that the female victim who was first beaten bloody then suffocated and strangled to death with her own underwear is somehow morally superior to the woman who sits in the police station and calmly explains to officers why she found it necessary to shoot to death a rape/burglary suspect in her own kitchen.

In the former case, the (dead) victim is lauded as a "role model." Taking no precautions and then meekly surrendering to criminal violence, while doing nothing to protect yourself, is today touted as one's ultimate civic duty.

In the latter case, the woman is mercilessly rebuked for "taking the law into her own hands," and "trying to do our job." Newspaper editors and media anchors predictably join the feeding frenzy. Of course, when the police arrive too late, the mayor meekly explains that "Police can't be everywhere now, can they?" Therefore, the only way she can gain their approval is by being a "good victim."

Ultimately, self-deception, even when encouraged by "mainstream" society, is still delusional. As warriors, we must always look down upon our enemy with disdain, never up at him in fear. We must never doubt our own magnificence and thus our right to defend ourselves (by any means necessary) even when that notion is disparaged by the weakling majority. We must not cut our conscience to fit this year's fashion! A man more right than ten of his neighbors makes a "majority of one."

South Africa

4 Aug 2003

Unhappy story from SA, from a friend and student there:

"One of my (and, by proxy, your) students owned a security company here that specialized in high-risk movement of cash ("AIT," or Assets in Transit). Rudolph trained extensively and had taken five courses with me. On the job, he always wore body armor and two pistols.

Last week, Rudolph went to collect money from a local Pick & Pay store here in Johannesburg. He arrived early in the morning, went to the back door, and unlocked the security door that protects the inner door. He entered and locked both doors behind him, as per standard procedure.

Once inside, he collected cash boxes from the staff. Then, someone knocked on the outer security door. Rudolph, apparently thinking it was one of his staff, opened both doors. He was greeted by six armed robbery suspects, three with Kalashnikovs and three with pistols.

The robbers quickly forced their way inside, making everybody lie, face down. Then, they searched each individual for keys to the cash boxes. When they got to Rudolph, they disarmed him and, upon seeing that he wore body armor, obviously presumed him to be a threat. They immediately shot him in the back of the head with a pistol and twice through his soft body armor in the back with an AK. They then picked up the cash boxes and escaped. None of the others there were harmed. Suspects are still at large."

Lesson: You can do everything right a thousand times, then let your guard down only once. In Africa, few get second chances. Desperate, determined, and wholly evil men lurk everywhere.

Sometimes, we all may be faced with this formidable choice: I can die fighting here and now, or I can be murdered on my knees after I have been rendered helpless. In a case like that, you may as well go for it, because you're probably going to die anyway. Everyone needs to think about this unhappy situation in advance. For my part, they're not getting me without a fight!

28 Oct 2003

Local police are under attack not only in Iraq. This is from a friend in SA:

"Last week, Mr. Selebe, our Minister of Police, requested the public to come forward to assist investigators with information on criminals who target police officers. We're suffered a rash of attacks lately. Most have been fatal to the targeted police officer(s).

As if to answer his plea, the next day two men walked into a police station here in Capetown asking for directions. Without warning, one pulled a pistol (type and caliber unknown) and shot the constable who was trying to assist them. He died at the scene. Two other constables in the station at the time returned fire, but to no effect. The two suspects fled on foot, apparently uninjured. They are still at large.

I do not wish to run our guys down, but they are inexcusably undertrained and thus routinely panic and default to 'spray and pray.' The average constable attends a range exercise only once a year and fires fewer than a dozen rounds."

Lesson: Resources expended on intensive and frequent training sessions pay big dividends at those times described above. Isn't it curious how we never seem to be able to find the money to train, but we always seem to be able to find the money for elaborate funerals?

My friend and colleague, Tom Givens, recently noted that crime statistics are all rubbish. In fact, "security" is merely a word we've invented for the sole purpose of deceiving ourselves. It is used to describe a circumstance that doesn't exist! There is no such thing as "high crime" or "low crime." Those terms have meaning only to statisticians. For each of us individually, crime is either 100% (when we're in the middle of incident such as this one), or it is zero (for the time being). In the face of daily news, only naive fools remain unprepared and unequipped. Sage officers (and others) will seek out training rather than wait for it to be provided. At the moment of truth, you'll be on your own, and there are no "degrees of dead."

29 Oct 2003

Comments on the SA shooting from a friend with a large, domestic PD:

"Sounds like our PD! Our only requirement for qualification is thirty rounds annually. After you chase down the stragglers who are actively avoiding qualification, because they are poor shooters and don't like the associated embarrassment, it can be as long as sixteen months in between qualifications. As a result, our shooting statistics are not a source of joy:

OIS statistics (generated by the FBI) indicate a national average for hit percentages of 18%. Ours is an impressive 6%.

Another statistic you will find depressing relates to lack of weapons preventive maintenance. Last year, during qualifications, five hundred handguns were immediately deadlined by our armorers! They were conspicuously unserviceable upon initial presentation or broke during the thirty-round exercise. These guns were from our patrol officers who came to the range during their tour of duty. Most involved officers had not the foggiest idea how long they had been carrying their pistol in that condition. I wonder what these officers' partners were thinking when they realized that they had been working with someone who has been carrying a 'dead man's gun' for weeks or even months.

It is scary and discouraging for those of us who are trying to keep these guys and gals trained and ready to go. But, we press on!"

Comment: Unfortunately, what is described above is not unusual for a big police department where officers are considered "expendable" by politicians and bureaucrats alike. Thank heaven for dedicated trainers like my friend here, who daily fight apathy from both directions. They do more good than they know!

25. Terrorism

3 July 2003

This is from a young military officer in the Israeli Army:

"Israeli military training is now focused exclusively on small unit tactics, urban warfare, and crowd control. We all hope that IDF's current obsession with Intifada will not cause us to forget 'combined arms' skills we will need to fight Syrians, et al, in yet another conventional war. I doubt that the Syrians will keep us waiting much longer!

Our reputation for innovation and improvisation notwithstanding, we are still teaching the 'chamber empty' method for carrying pistols. It an obsolete method left over from the 1940s, when many of the pistols carried by our nation's founders were broken-down discards and were not drop-safe. I don't know why we can't finally shake this outdated practice, but it continues to be taught.

We are currently teaching both military and civilians gun carriers that 'once he is down, keep him down.' That is, we are to rapidly close and put two rounds in to the head of the wounded, would-be terrorist. We've decided that we only want to fight these bastards once!

The locally made Galil rifle has been relegated to rear-area defense. Front-line troops are carrying American M16s, most with optical sights. The Galil is a great rifle, but it suffers from a perceived weakness of the Kalashnikov system, namely the difficulty of attaching optics.

There is another reason: at the end of the day, it came down to 'why should we continue to build these expensive weapons, when we can get M16s from the Americans for free?'"

Comment: The Israelis are great people, but they suffer from the same species of self-deception we do. Specifically, that there is a mechanical device out there somewhere that will adequately substitute for personal competence and common sense.

15 July 2003

The latest on Homeland "Security" from a student who works at a military arsenal:

"Those assigned to 'security' have been instructed to carry their M9 (Beretta 92F) pistols with an empty chamber. When I expressed concern and confusion over this procedure, our colonel said that it was 'safer to have the weapons with an empty chamber, because of the (obviously poor) level of training of the security force.' When I was asked why we were not better trained, I was told that the Army 'doesn't have the time or resources,' (nor, obviously, the interest)!

It gets better: We have been told that we are no longer going to be allowed to carry high-performance, hollow point ammunition in our M9s. This same idiot colonel is recommending hardball. I, once more, asked about lack of immediate incapacitation and the simultaneous risk of over-penetration, as our facility has many tanks filled with toxic chemicals. The colonel admitted that he had not even thought about the over-penetration issue, then insisted that hardball was still 'the best way to go.' I pointed out that no civilian police agency carries hardball anymore. The colonel expressed surprise but would still not be moved by facts.

Finally, we have been told that our shotguns are being returned to the armory, because 'all we need here is small arms.' I reminded the colonel that shotguns ARE 'small arms.'

So, we now have lightly armed, marginally trained 'security' officers, firing ineffective, non-expanding ammunition in a supposedly critical military facility with tanks full of acids and explosive gases!

Again, the difference between the high-sounding, presidential rhetoric on Homeland 'Security' and the pathetic reality down here where the rubber meets the road it stark and frightening. In a real attack, most of us would be killed in the first minute. We are nothing but cannon fodder!"

Comment: Here is another example. We in the USA have elevated self-deception to an art form! I blame politicians, of course, including our current Commander in Chief, for naively believing what he is being told by his gaggle of atta-boy generals instead of seeing for himself. However, the real villain in this story is the colonel mentioned above, and thousands like him. He may be a West Point pretty boy, but he is also a pathetic coward! He knows full well his 'security' is a joke and the lives of his men are at risk, but he is so comfortable he doesn't want anything to ever change. He has a chance to be a hero, but he would rather be an atta-boy. Shame on him!

28 July 2003

Heartening comments from a friend in FRG (Germany):

"Please accept my personal apology that my country, or rather our 'leadership' hasn't supported the USA in fighting for liberty. Many here are likewise inclined. It's shameful indeed, having enjoyed the USA's protection against Communism for the last fifty years, and now, when we have the chance, giving nothing in return. Unfortunately, our country is going down the road with the Socialist-Greens for three more years."

Comment: There are still many good people in the world. Unhappily, good people are not attracted to politics, and leftist politics only attracts the sleazy and amoral!

12 Nov 2003

The view from Saudi Arabia, from a friend who works there:

"I had a visit with my superintendent yesterday, after work. He has always been a pretty good source for local opinion. He is a typical government guy. His English is thick, but understandable. Like everyone here, he has been educated from birth to harbor a vile, rabid hatred for Western Civilization and everything associated with it, and, like most here, he is convinced that Saudis are God's chosen people and that the world would be far better off if all Christians and Jews suddenly found another planet!

However, in view of recent events, this was the first time we ever met that he didn't hit me with some issue about the US presence in Iraq. He didn't even dump on my president, and he didn't say (as he always has in the past), "I look at this from a religious perspective…."

He is scared! I've never seen him like this. Clearly, he cannot understand the reason for his beloved Al Queda striking the Arabian compound this week, killing and maiming all kinds of Arabs. What is going on here is, of course, an insurgency. It is crystal clear to all us Westerners, but this Saudi and his countrymen don't want to face the possibility of such a thing. It is just too big for any of them to imagine.

Up until now, police and military training here has been mostly a game (like everything else here). When they conduct a raid, they typically send out twelve cops and bring back seven. Not a recipe for long term success! In fact, most government security forces in the Gulf are sweating the day when this comes to their jurisdiction!"

Comment: We may live to see the Saud family topple in Saudi Arabia! Ten years ago, they seemed impregnable. But, they danced with the devil, and now the devil wants to lead. There is a lesson here!

Part 5: History

26. America

20 Dec 2002

The turning points: Bunker Hill, Saturday, 17 June 1775, and Saratoga, Friday, 17 Oct 1777, two years and four months later.

During the American Revolutionary War, the Battle of Bunker Hill showed everyone that American militiamen could stand up to British regulars, even though British occupied the disputed real estate at the end of the day. Prior to that, everyone believed Americans would all run away instead of fighting. The Battle of Saratoga showed everyone that Americans could not only stand up to the British but could actually win a decisive victory and even capture a British crown officer ("Gentleman Johnny" Burgoyne) in the process. It was this battle that persuaded the king of France to throw in with the Americans. Holland and Spain quickly followed. From that point forward England's attention was divided, and all the American Colonies had to do was hold out and give their army minimal support, which is about all they did!

Aside from the Battle of Saratoga and a few others (such as Washington's surprise victory at Trenton in December of 1776), the American Revolutionary War featured few clear victories, even fewer decisive ones. The famous British naval blockade that was so effective with other British colonies, was useless here. Like most wars, this one gradually fizzled out with the two sides coming to an unworkable agreement to which neither had any intention of adhering. No sooner was the "Peace of Paris" signed in 1783 than British inspired and led Indian insurrections began to plague the American western frontier. Called the Northwest Wars, it started with St Clair's ignominious trouncing at the hands of Little Turtle in November of 1791 and ended with "Mad" Anthony Wayne's victory over Little Turtle and Tecumseh at Fallen Timbers on 20 Aug 1794.

The British were still smarting when fighting broke out with Napoleon in 1803. Sea battles in particular became ruthless, with the British contemptuously refusing to acknowledge American citizenship (which they had agreed to do) as they forcibly "recruited" sailors. The smoldering relationship between England and America ignited into open conflict once more. In August of 1814, amphibiously landed British troops brushed aside American defenders, entered Washington, DC, and burned many government buildings, including the White House, before leaving. President Madison steadfastly refused to negotiate, and, after an astounding mauling at the hands of American soldiers in Baltimore a month later, British troops got back on their ships and withdrew. The fight then moved to the City of New Orleans where, in January of 1815, the British contingent of over two thousand, under Pakenham, were wiped out in a single morning by Andrew Jackson. Jackson's stunning victory provided a moral boost to the Americans, but it actually took place a month after the war had officially ended with the Treaty of Ghent, Belgium on 24 December 1814.

In 1818, then in Florida, Jackson executed two Englishmen (one was hanged; one was shot) who were advising local Seminoles. These British nationals were the last two casualties of the protracted "Revolutionary War," America's longest war, which had begun forty-three years earlier at Lexington Common in Massachusetts. American and British soldiers would never fight each other again.

Back to our story:

In June of 1755, during the French and Indian War, a young Thomas Gage served with Braddock at the infamous Battle of Monongahela and was wounded. In fact, he narrowly escaped death, along with two other notables, Daniel Boone and George Washington. By 1773, Gage, was a general in the British Army, had married an American woman from New Jersey, had spent most of his adult life in the American Colonies, and was indeed planning on retiring in America rather than returning to England (as things worked out, he became governor of Canada)

In 1773 Thomas Gage, in fact, found himself commander of all British troops in North America. His army was top heavy, with an excess of semi-retired colonels left over from the French and Indian War. In addition, his army had amply demonstrated its ineptness at fighting Indians, but Gage was confident he could keep the lid on his fellow Englishmen. Unfortunately, He consistently overestimated British loyalty among colonists and underestimated the ability of colonists to organize an operative fighting force that was not afraid of his "regulars."

With the Boston Tea Party in December of 1773 and the subsequent "Intolerable Acts" which King George III used as a reprisal, local unrest was getting out of hand. Gage had an extensive spy network, and he knew a colonial shadow government, complete with its own "army" of Minutemen, was in place and operating. They were well armed and had military equipment stored in various locations.

Trouble was brewing!

On Wednesday, 19 Apr 1775, Gage's troops approached Lexington Common in search of military supplies that Gage's spies had told him were stored there. The rebels had their own spy network and knew of the British approach. Captain John Parker, commander of the local contingent of Minutemen, ordered his hastily assembled men to "Stand your ground. Don't fire unless fired upon. But, if they want a war, let it begin here."

"Lay down your arms, you dammed rebels, and disburse," came the command from Major Pitcarin of the Royal Marines. Parker told his men to disburse, but to keep their arms. A shot was fired, and Pitcarin's men immediately fired a volley into the Minutemen. The volley was followed with a bayonet charge. One British soldier was wounded. Eight Minutemen were killed and as many wounded. The rest ran away. Pitcarin's unit went on to do what it had come to do. They located military stores, and, without even bothering to break open the crates and boxes, threw them all into a local pond. They were later recovered, intact, by the Minutemen.

That was that, Pitcarin thought, and his men started marching on toward Concord to finish their job. At Concord, they met more organized resistance, and street fighting resulted in a number of dead British soldiers. Colonel Francis Smith, Pitcarin's commander, decided to retreat back to Boston before more of his men were killed or wounded.

But, word of the clash had spread, and Minutemen from all over rushed to the area and began to line the road back to Boston. Many had rifles instead of muskets. They fired on the British from long ranges using stone fences, trees, and logs as cover. Volleys fired in return were ineffective, as were bayonet charges. Smith's entire unit, thoroughly chewed up, was, in fact, about to crumble as it crossed Charlestown Neck to Boston and safety. Smith's losses were seventy-three killed and several hundred wounded. He and Pitcarin were in shock.

The American Revolutionary War had well and truly begun!

American militia units were everywhere mobilizing. London demanded quick and decisive action from "Blundering Tommy" Gage, as he was coming to be known. They sent Gage more men and three new generals, Howe, Clinton, and Burgoyne. Howe and Clinton were to suffer terrible losses at Bunker Hill. Burgoyne would be captured at Saratoga, after seeing his entire army decimated by Daniel Morgan's riflemen.

On the evening of 16 June 1775, Militia colonels Israel Putnam and William Prescott led twelve hundred armed farmers to Breed's Hill in order to fortify it. Bunker Hill, just behind it, would serve as a secondary position. The hills were strategically important to Boston Harbor, and Gage knew well the importance of driving the rebels out. General Howe, right up with his men, engineered three determined charges. All failed, with catastrophic losses. Defensive positions, dug by the farmers were well designed and held, and the rebels were able to reload faster than Howe had thought possible. In addition, farmers equipped with rifles were tasked with picking off British officers and NCOs, and they performed their duty with fearful precision! British ships shelled the hill, but with little effect. Howe organized several more charges, all with the same result. Seeing the wholesale carnage all around him affected Howe greatly. He never recovered.

Rebels in their revetments were nearly out of ammunition. Nails, bits of broken glass, and pebbles were substituting for lead balls. Powder was nearly gone. A final British assault broke through, but the Americans retreated in good order and most escaped. British troops finally occupied the two hills. Clinton was heard to say, "A dear-bought victory; another such would have ruined us!"

After Bunker Hill, even formally loyalist Georgia and New York joined the rebellion. The point of no return was now well past.

Howe replaced Gage but was never his daring self again. He failed to attack George Washington in Long Island and later during the Battle of Brooklyn, allowing Washington to escape both times. He failed to attack at White Plaines. He even failed to attack the starving American Army at Valley Forge during the winter of 1777, which would have been an easy victory. Howe had no stomach for it anymore.

By the time Howe was replaced by Cornwallis, the war was mostly over. Cornwallis ran his army ragged chasing rebels throughout the Southern colonies, only to stagger into Yorktown, out of ammunition, unable to care for his wounded, and starving. It was there he met his final defeat at the hands of George Washington and Alexander Hamilton on 17 October 1780. Two more years of sporadic fighting went by before the Peace of Paris was finally signed on 3 February 1783.

Lesson: If that bedraggled group of armed citizens at Bunker Hill had broken and ran, the revolution would have been all over at that point, and we would still be a British colony. But they stood up and held their ground. Sometimes we must be heroes, whether we like it or not! "Difficulty" is an excuse history never accepts.

24 Apr 2003

Hobnobbing with gangsters:

Dr Reinhardt H Schwimmer was a successful young Chicago optometrist who apparently found associating with local gangsters exciting. It would be his undoing.

In early 1929, Al Capone had entrusted the job of eliminating his archrival in the illegal booze business, George "Bugs" Moran, to his ruthless, but sharp-witted protegee, "Machinegun" Jack McGurn. An elaborate ambush was set up, with Al himself conveniently in Florida. At one of Moran's camouflaged liquor depots on Chicago's north Clark St, under the front "SCM Cartage Co," McGurn's scouts (hired from Detroit's Purple Gang), from a boarding house across the street, spotted Moran and his crew enter the building.

What looked like a Chicago Police car pulled up in front of the building and four men emerged, two of whom were wearing CPD uniforms. The other two were in heavy overcoats. The unformed men burst into the building and ordered everyone to turn around, face the wall, and get their hands up. Expecting the customary shakedown and subsequent payoff, none of Moran's men were particularly concerned for their safety. Immediately, the two men in overcoats produced Thompson submachine guns and shot to death all seven of Moran's hapless associates, including the astonished Dr Schwimmer. Seventy expended 45ACP cases were found at the crime scene, indicating one of the Thompsons probably had a fifty-round drum, while the other had a twenty-round stick.

Actually, the scouts were mistaken. Moran himself arrived late and quickly left without entering the building when he noted suspicious activity. Of the seven murdered in the subsequent St Valentine's Day Massacre on Thursday, 14 Feb 1929, Moran himself, the one Capone really wanted, was not included.

Several of the alleged shooters were eventually arrested, but none were ever convicted. Capone himself was also never convicted for any act of violence. In 1932, he ended up in California's infamous Alcatraz Federal Prison anyway, but on tax-evasion charges. During his internment, his mental faculties deteriorated catastrophically due to a lifelong case of venereal disease which he probably contracted in his early years when working as a bouncer in a sleazy cathouse in NYC. When he was released in 1939, he was a broken, dilapidated, weak-minded invalid. He only lived another eight years, all in seclusion.

Moran would also end up in prison, in his case on robbery charges in Ohio. He died in prison, outliving Capone by ten years.

Lesson: Patterns of chronic criminal behavior are acquired early in life and are seldom cast off. Habitual criminals could not lead decent lives, even if they wanted to. At the appropriate moment, even "respectable" criminals will predictably revert to type. They are always capable of fearsome violence. Associating with them on any level is foolish and reckless, as the good doctor (belatedly) found out!

7 July 2003

Shays's Rebellion and the Second Amendment:

The series of conflicts that today are blended together into what historians call the French and Indian Wars in the mid-1700s convinced the French that their national interests in North America could be adequately defended by their Indian allies. Therefore, France, who at the time had the largest land army in Europe, sent no troops. The British, equally convinced that their Indian allies were duplicitous and unreliable, sent over troops by the boatload! The future of what would become the United States of America was thus decided. In the later half of the Eighteenth Century, French influence in North America dwindled, as British influence expanded.

Expanded too much, as it turns out. American colonists eventually threw off characteristically heavy-handed British rule. The Revolutionary War, which started at Lexington and Concord, Massachusetts in 1776, essentially ended at Yorktown, Virginia in October of 1780, although an official peace treaty was not sighed until 1783, and hostilities did not end until 1818!

Guns & Warriors: DTI Quips – Volume 2

At War's official end, having done his job, General George Washington retired to his farm in Virginia, having no further interest in military operations and even less in politics. He would probably have lived out his days there in quiet seclusion as a gentleman farmer were it not for a decorated Revolutionary War veteran in rural Massachusetts named Daniel Shays.

Under the Articles of Confederation, the only document that bound together the newly independent thirteen (former) colonies, each colony was now, in essence, and independent country. In Massachusetts, politicians (then, as now) saw it as their duty to tax people to death while forging political alliances with powerful interests, all of which were located in Boston, the only big city in the colony. Led by Governor James Bowdoin, a campaign to foreclose on debtor farmers in rural Massachusetts in an effort to seize vast tracts of private land (which could then be doled out to political supporters) went on with scant notice outside the Massachusetts Colony itself. Rural towns everywhere pleaded with the Assembly in Boston to address their plight but were ignored.

Massachusetts citizens began wondering why they had bothered to fight a war against the British. It struck them that the "Revolution" had failed them, that they had merely exchanged one tyrant for another. They had been led to believe that a democracy was a virtual guarantee against tyranny. Now, it was obvious that tyranny could exist, indeed flourish, even within a democracy. Rural areas could not muster enough votes to counter the vast reserve of votes in population centers, like Boston. In Boston, groups and constituencies could easily be bought and manipulated, so politicians seldom bothered to venture forth beyond the city limits (much as is the case today).

Daniel Shays and his wife Abigail were farmers in rural, western Massachusetts. Shays never allowed a portrait of himself to be painted, so we don't know what he looked like. We do know his service during the War was characterized by conspicuous bravely and devotion to his country.

Seeing no sympathy in their plight emanating from Boston, armed insurgents in the Berkshire Hills and the Connecticut Valley, under the leadership of Daniel Shays and others, began in August of 1786 to forcibly prevent county courts from sitting to make judgments on foreclosures. In fact, in September of that year they forced the state Supreme Court at Springfield to adjourn before it could conduct any business. The Boston elite finally took notice, but, when they tried to enlist the help of the state militia, they found it of scant assistance as most of its members, particularly Revolutionary War veterans, sympathized with Shays and his followers.

Like most who participated in the 1786/87 Shays's Rebellion in central and western Massachusetts, Shays was a noble and upright citizen, with no criminal history. None of this interested Bowdoin (who was not a veteran), and he ruthlessly used his influence with the press to portray Shays and his colleagues as sleazy, threadbare, trailer trash. That false portrayal still persists today.

On 25 January 1787, a poorly organized armed group, led by Shays, marched on the US arsenal at Springfield in an effort to break in and equip itself with modern, military arms. They were repulsed by a hastily assembled state militia unit under the command of Major General William Shepard, who had arrived in the nick of time, as no federal units were available to protect the facility. Shepard used the arsenal's cannons on the rebels and killed four. Five times that number were wounded. The assault subsequently fell apart, and the rebels fled in disarray.

They were pursued by General Benjamin Lincoln, leading another hastily assembled army, to the town of Petersham. There, early in the morning of 4 February, Lincoln surprised Shays and his men, but the weather was so bad that a full-scale attack could not be mounted. Most of the rebels, including Shays himself, escaped.

On Tuesday, 27 February, Colonel John Ashley routed a group of insurgents, under the command of Perez Hamlin, near Sheffield, killing two and wounding thirty. A hundred prisoners were taken. The 27 February engagement is considered by most to have marked the end of the

Rebellion. However, sporadic activity continued for months afterward before dying out altogether by the end of June. The Sheffield battlefield is today marked by a single, stone monument, the only reminder of Shays's Rebellion to have survived to the present time.

Most of the rebels were promptly pardoned. A number in the leadership group were sentenced to be hanged, but, in the end, only two were. With an election coming up, Bowdoin was anxious to put it all behind him with a minimum of fanfare, knowing the Rebellion still garnered a good deal of sympathy. Bowdoin was soundly defeated anyway, and his successor, John Hancock, equally anxious to move on, pardoned nearly all who had not been pardoned before.

Daniel Shays fled Massachusetts and eventually settled in western New York, although he himself was pardoned by Hancock in June of 1788. He died in New York in 1825, then in his seventies, of natural causes, outliving both Bowdoin and Lincoln by many years. In New York, he was noted for his aversion to politics!

Shay's Rebellion scared the snot out of the governors of the other twelve colonies and literally drove them to a Constitutional Convention in Philadelphia in 1788, something for which most had no enthusiasm previously. It also drove George Washington out of retirement and ultimately into the office of president. A strong, central, federal government was now seen as a crucial necessity. In fact, opponents of the new Constitution were incessantly accused of being sympathetic with Shays and his rebels.

However, in the middle of the convention, James Madison and Elbridge Gerry insisted on a Bill of Rights, fearing "Massachusetts-style" repression would be a consequence of a concentration of power at the national level. With Shays's Rebellion still fresh in their collective memories, the notion instantly struck a resonant chord. In the end, it became obvious that a constitution, without amendments protecting individual rights, would not be ratified. In December of 1791, the new Constitution, with a Bill of Rights, was finally approved by all thirteen colonies (now states) and took effect.

Without Daniel Shays and his audacious Rebellion, there probably never would have been a Constitutional Convention. The colonies would have remained disunited and would have been ultimately reabsorbed, one by one, back into the British Empire.

Without Daniel Shays, George Washington would have probably stayed in retirement and not become our first president.

Most importantly, without Daniel Shays, Thomas Jefferson probably never would have expressed the opinion that an occasional armed insurrection is healthy for a democracy. Armed protest against tyranny would probably not have thus imbedded itself so thoroughly into American democratic thought. Indeed, the Second Amendment was included in the Bill of Rights precisely to support that conviction.

Without Shays, there would probably never have been a Whisky Rebellion (in western Pennsylvania in 1794), nor an American Civil War (seventy years later).

Comment: Like so many heroes of human history, Daniel Shays is scarcely known today. He was a quiet man of principle who never wanted to be a celebrity. Like most decent and honorable men, he was never attracted to politics. But, when history called upon him, he went forth boldly and did what clearly had to be done. Good show, Dan!

28 Aug 2003

Pickett's Charge at the Battle of Gettysburg, PA on 3 July 1863 and Rawlinson's Charge at the Battle of the Somme River in France, 1 July 1916. Same tactics. Same result. The two catastrophes separated by only fifty-three years, almost to the day!

General Robert E. Lee had attempted to roll up Union lines on Cemetery Ridge south of Gettysburg from both ends. Unfortunately, his best generals were nearly all dead by the summer of 1863 (most notably, Tom

"Stonewall" Jackson, killed at Chancellorsville), and both attempts had failed by mere inches, but failed, nonetheless. Now, he had no choice but to attack the Union center or withdraw from the battlefield in disgrace. It would be a long and desperate rush over open ground, but he had the best artillery in the business, and his chief of artillery had graduated at the top of his West Point Class. Lee convinced himself that his artillery would dislodge and disorganize Union lines, paving the way for a decisive and crushing bayonet charge. With his army thus conclusively routed in their own territory, the latest in a long and dreary litany of thrashings at Lee's hands, President Lincoln would have no choice but to negotiate. Murderous fighting, that so characterized this conflict, would finally come to an end!

Fifty-three years later in France, British General Rawlinson was also a man under pressure. The "Great War," that had been welcomed by so many as a "great patriotic adventure," had since 1914, deteriorated into a stagnant, self-perpetuating, pitiless massacre that showed no signs of ending. Trenches in Western Europe had stagnated. Commanders repeatedly persuaded themselves that, with just "one more push," German lines could be broken, a breakthrough could be established, and they could then rush to the rear. Months came and went, and no breakthrough happened, despite innumerable attempts. But, Rawlinson had artillery, lots of it, and lots of shells, many made in America. Like Lee, he was convinced that his artillery could not only drive Germans from their trenches, but that it could also cut through and blow away the endless layers of barbed wire that protected them. German barbed wire was so thick, it was said light could not shine through it. With Germans all killed or driven away and barbed wire gone, his infantry could then casually walk across no-man's land and drive to the German rear. The long-awaited breakthrough would finally be at hand!

Both charges failed catastrophically! Both wars would go on for two more murderous years. Both generals could not bring themselves to confront the fact that their vaunted artillery bombardment had failed to do what it was intended to do.

At Gettysburg, exhausted Confederate artillerymen sent most of their projectiles over front line Union trenches. Rear areas were thus decimated, but the front lines came through the barrage largely intact. Confederate infantry, along a two-mile front, were subsequently wiped out. The few that made it to Union lines were killed or captured there. The Union line held, and the remnants of Lee's infantry staggered all the way back to their starting point on Seminary Ridge. Lee was forced to withdraw from Gettysburg and salvage what remained of his army. George Meade, the opposing Union commander, failed to pursue, and Lee escaped to fight once more, but the outcome of the war was now certain.

Rawlinson's artillery also did not deliver the miracle he had been assured it would. One-third of the 1.5 million shells fired over the five-day artillery preparation never went off at impact. German infantry positions were far deeper and stouter than anyone thought, and German barbed wire was mostly unscathed by the bombardment. Rawlinson's own patrols had told him this, but, since that is not what he wanted to hear, the information was brushed aside.

German machine gunners quickly manned their Maxim Guns as soon as the bombardment shifted to the rear. British infantry units, on a fifteen-mile front, snaked through gaps in their own barbed wire and reassembled on the opposite side in smart ranks, facing no-man's land. They were then scythed down like so much wheat by German machine guns. The slaughter went on all day, with few British soldiers advanced more than a few hundred meters. The Battle of the Somme was to last 140 days, but the first day alone would see the deaths of 20,000 young British soldiers. Three times that number were wounded. An entire generation of British youth were wiped out. It was to be the largest loss of life ever in a single day of fighting, before or since.

Lesson: Any time you are told of miracles, miracle weapons, miracle ammunition, miracle fighting techniques, etc, never allow yourself to be taken in. Never be reluctant to face facts. The enemy may not be impressed with your "superior technology." When your bluff is called, you had better not be bluffing!

12 Sept 2003

Hard Luck Ambrose, July 1864

Irwin McDowell had made a hash of things at the First Battle of Manassas (Bull Run) in July of 1861, sending the signal to everyone that what would eventually become known as the War Between the States would well and truly be a bona fide war, and that it would last longer and claim far more lives than anyone thought it possibly could. President Lincoln immediately sacked McDowell for McClelland.

Sarcastically called "the Virginia Creeper," McClelland appeared to Lincoln and the rest of the nation to be moving too slow against the rapidly organizing Confederates. Lincoln finally lost patience with McClelland's slow progress and sacked him in favor of affable and well liked Ambrose Burnside. Burnside himself warned Lincoln that he "was not competent to lead such a large army," but he got the job anyway. Lincoln would live to regret not taking Burnside's own advice- twice!

Burnside's signature beard/mustache combination gave him instant recognition, as the increasingly desperate future of the Union fell unceremoniously upon his shoulders. The modern term, "sideburns," comes directly from Burnside's name.

Burnside promptly lived up to his own low expectations of himself at the Battle of Fredericksburg, VA in December of 1862, where he directed one pointless charge after another against entrenched Rebels on Marye's Heights. His foolish charges all failed, and Burnside himself became so despondent upon seeing the purposeless slaughter that he decided to lead the last charge himself, which amounted to a death wish. He was ultimately talked out of it by his staff, because they would have all been killed too!

With his virtually leaderless army now on the verge of mutiny, an exasperated Lincoln sacked Burnside for Joe Hooker (who would, in turn, be outwitted by Robert E Lee and Tom Jackson at Chancellorsville and then himself be superseded by John Pope). Burnside, thoroughly embarrassed and discredited, was subsequently shuffled to the rear and should have been shuffled out of the army altogether, but he kept his rank and stayed in the fight. Two years later, he would, one last time, lend his own special brand of blundering incompetence to the Union war effort.

The twin (and nearly simultaneous) disasters at Vicksburg, MS and Gettysburg, PA in July of 1863 had marked the beginning of the end for The Rebellion. There was now no chance of foreign intervention by the British, and Lee was fighting what amounted to a spirited retreat, but he had lost none of his tactical genius and still had some surprises up his sleeve!

The assault on Richmond, VA in July of 1864 (part of Ulysses S Grant's "Overland Campaign") lead by George Meade (now a subordinate, under Grant), featured a resurrected Ambrose Burnside in command of the Union IX Corps. Directly in Burnside's path of advance was an entrenched Confederate position at Petersburg. Colonel Henry Pleasants, one of Burnside's subordinates and a former mining engineer, told Burnside that he could tunnel under the Confederate positions, emplace explosives there, and then blast a gaping hole in their line. Burnside's men could then rush to the rear, and all the way to Richmond, in a stunning breakthrough. The end of this terrible War would finally be at hand.

Burnside was skeptical at first, but Pleasants was an excellent salesman as well as a gifted engineer, and Burnside was finally persuaded to give the plan a try. Even Grant, fresh from his own disaster at Cold Harbor, gave the go-ahead, although his staff remained skeptical. Pleasants and his men dug the tunnel with enthusiasm and expertise, overcoming all obstacles. When it was completed, they packed it with several tons of black powder, more black powder than Pleasants, nor any of his men, had ever seen before in one place. Pleasants didn't know how big a bang it would make, but he was confident it would be adequate. It was far more than "adequate!" In fact, the explosion was so big that it made an enormous. asteroid-like crater. A gap was blown in the line of the astonished rebels to be sure, but the blast was so stunning, many of Burnside's own troops fled to the rear in terror!

The follow-up infantry charge was slow and disorganized, because Burnside himself was far to the rear hiding in a bunker, afraid and having apparently lost interest in the whole operation. The commander of the assault troops, General Ledlie (a train engineer by profession), a pathetic alcoholic, was drunk and incoherent! As soon as the charge commenced, he also fled to the rear.

As a result, Union assault troops quickly disintegrated into little more that a horde of tourists. They should have simply gone around the crater and brushed aside the remaining confederate defenders, but, without leadership, they charged pell mell into the crater itself. The crater seemed to have a mesmerizing effect, as wave after wave of leaderless Union troops, nonchalantly sent forward by a detached Burnside, were attracted into it. Most of them died there in a slaughter worse than Fredericksburg!

Confederate defenders, stunned but still expertly led, immediately assembled on the far ridge of the crater and began firing down into the hapless and utterly disorganized Union throng, who were by now milling about aimlessly. Confederate reinforcements, under General Mahone, soon arrived and turned the defeat into a complete rout. A Confederate commander, struck by the pitiful and lopsided slaughter, finally shouted at the few Union troopers who remained alive and said, "Why don't you surrender?" A lone Union trooper replied, "Why don't you let us?" Then, in their first and only display of coordinated action that day, Union troops all simultaneously dropped their weapons and put their hands up. Shooting stopped. Union troops were marched off as prisoners, although many black soldiers (several black units were involved in the assault) were shot on the spot by enraged Southerners.

Meade finally got through to Burnside and ordered him to stop the "advance," salvage as much as he could, and retreat. Grant described the fiasco as, "the saddest affair I have ever witnessed." The War, which could have effectively ended that day, would now go on for another nine months.

Burnside was fired, for good this time. He would never command troops again. He went back to his native Maryland where he eventually became governor and later a US senator. With his trademark beard now white, he tried (in vain) until the day he died to resurrect his well-known reputation as the incompetent bungler who was directly responsible for countless unnecessary casualties.

On the positive side, Burnside became an outspoken advocate for veterans and was one of the Founders of the NRA, and its first president!

After the War, Pleasants went back to the mining industry. Ledlie was court-martialed, slapped on the wrist, and went back to the railroad.

Lesson: It takes a special species of incompetence to snatch defeat from the jaws of victory. Pleasants was a genius and a hero, but, teamed with ditherers, atta-boys, and drunks, his brilliant plan still failed. Ultimately, the blame for putting unqualified people, people who had painfully proven their incompetence, into important positions, falls on Lincoln. He should have known better. Thousands paid the price for his poor decision- twice!

27. The World Wars

30 Mar 2003

The Forgotten Counterattack

The Normandy Invasion of June of 1944 had been a success, and by August the war was going favorably for the Allies. Americans were moving toward Paris. Patton's Third Army was making headlines with lightning advances across Brittany. Even British General Montgomery was (slowly, as always) beginning to move at Caen. Operation Cobra had succeeded in capturing St Lo, despite two separate instances of miss-bombings by friendly aircraft.

The 30th Infantry Division had then taken the lead at St Lo. On August 6th, they relieved the 1st Division at the small French town of Mortain. The 30th did little more than assume prepared positions.

The 30th had only 57mm cannon and a unit of towed, 75mm rifles for use against tanks. These antitank guns were the last units to emplace. However, many of the prepared gun positions they took over were unsuitable for these low-mounted, towed weapons. Everyone was tired, and scant attempt was made to improve the positions, as all expected to resume the attack momentarily at first light. As exhausted soldiers of the 30th took over defensive positions, they were told by the departing 26th, "There are no Germans here."

Having failed to stop the Normandy beach landings and Cobra breakout, German commanders were now desperate to halt the Americans, and Patton in particular. They calculated that the key to stopping Patton was to cut his supply lines. Recognizing this, Hitler, taking personal control, ordered a Panzer attack to cut through American supply routes. It was called "Operation Luttich."

On mostly open ground, the right front of the American line was manned by C Company. August 7th began for the 30th with the sounds of moving vehicles to their foggy front. Out of the fog suddenly appeared a column of Panzers. C Company was quickly overrun.

The 30th reeled in an attempt to absorb and eventually stop the German advance. Their antitank guns in their poor positions were destroyed, one-by-one, while having little effect on invading German tanks. Most German tanks and armored support vehicles were ultimately stopped by artillery, air strikes, and individual soldiers with 2.5" rocket launchers (Bazookas).

In an eerie prototype for the upcoming Battle of the Bulge, the German tank advance stalled after only five kilometers. Out of ammunition and out of gas, surviving German tanks ground to a halt. Much time was lost as Hitler personally meddled in the command structure. Seeing an opportunity to exploit this clumsy and ponderous German commitment, Omar Bradley masterminded a successful encirclement. It would be known as the "Falaise Pocket."

Many have asked why readers of WW II history seldom see nor hear accounts of the 30th. The reason is that the division commander, Gen. Leland Hobbs, hated the press. "Heroes," he said, "don't need 'coverage,' just ammunition."

Lessons:

Whenever you hear the words, "There is no enemy here," never believe it! When your position needs improving, do it now, while you still can.

When your survival plan depends on help from others, you might not make it. The help upon which you're depending may not materialize. Let your motto be, "Prepared for anything. Depending on nothing."

As in most things, overall victory comes from the heroic efforts of individual patriots, the names of whom have long since been lost in history. *"All this by gallant hearts is done. All this by patient hearts is born. And, they by whom the laurel's won ...are seldom they by whom it's worn."*

16 July 2003

The Dangers of Chronic Self-deception: Mussolini's "Invasion" of British-held Egypt, December 1940:

The World Wars

At the dawn of World War II, Benito Mussolini wistfully thought of himself as the modern incarnation of Julius Caesar. His ill-fated and pointless invasion of Greece in November of 1940 was already in full retreat, but he still longed for the center of the world stage for which he was convinced he was destined, particularly since Hitler, whom he considered a garish, but vulgar barbarian, seemed to be on a roll in Western Europe.

Since his 1936 invasion of Ethiopia had been a one-sided success, Mussolini decided that an invasion of Egypt from his bases in Libya would also be an easy victory and would, most importantly, get his name back on the front page. After all, British troops in Egypt had only armored cars, while his troops had tanks, and the desert was too hot for the British anyway.

Unhappily, the real situation on the ground was far different from Mussolini's "pretend version." In spite of numerous warnings from his field commander, General Graziani, about the substantial risks associated with going into Egypt, Mussolini continued in his dream world. If he had gone to Libya and seen the stark truth for himself, Mussolini would have personally witnessed the low moral, the broken-down logistics system that could never sustain any kind of invasion, and the utter lack of transport vehicles with which to get his troops to the front. His "army" in Libya was a joke! However, in the end, Graziani's warnings were brushed aside.

It turned out to be the shortest "invasion" in the history of modern warfare. By the time Mussolini's troops arrived, via foot march, at the Egyptian border, they were already starving and dying of thirst. Italian crews jumped out of their tanks and ran for the rear as soon as they saw the first contingent of British armored cars. Within weeks, nearly all Italian troops and tanks had been captured by the British. All surrendered with great enthusiasm. Few shots were fired. The whole "invasion" penetrated Egypt by only a few meters!

A German soldier was heard to comment to his British captors, "Next time, it's your turn to have the Italians!"

Comment: The foregoing is surely not unique in human history. At present, many of our national leaders suffer from the "Mussolini Syndrome." Concerned only with their own personal, political ambitions, they would rather "pretend" than face facts. They would rather listen to the soothing lies of their atta-boy subordinates than see for themselves what is really happening.

However, egoistic politicians can only pretend so long. Reality has a bad habit of crashing through the rhetoric veneer (as we see from the foregoing). That is when politicians also have to be good a finger pointing.

Heroes don't allow themselves to wallow in self-deception. They tell the truth, even when it is ugly and unpopular, and the shallow and self-centered hearers would rather "pretend." At the national level, there is currently a famine in the hero department!

28. Southeast Asia

18 Apr 2003

Massacre at Balangiga, Island of Samar, Philippines, Sunday, 29 Sept 1901

In the current Gulf War, we've seen desperate and despondent Iraqi fighters using what we consider deceitful and opprobrious tactics in a forlorn effort to drive out invading American soldiers. We have predictably reacted with shock and indignation. What we need to remember is that none of this is new, nor should any of it be unexpected. Despairing people will always find unconventional and revolting methods for opposing what they consider to be invaders, even though the "invaders" are actually liberators and rescuers.

During the Spanish-American War, young American soldiers found themselves in the faraway Philippines, once again attending to their nation's interests. Most Americans had never even heard of this part of the world before the war started. At that time, the Philippine Islands were a Spanish colony, and American Admiral George Dewey had already decisively defeated the Spanish naval fleet stationed at Manila Bay.

American soldiers encountered a few regular Spanish units on Philippine soil but were mostly opposed by loosely organized native Filipino guerrilla fighters who didn't like Americans any more than they liked the Spanish. Like today's Iraqis, Filipinos were poorly armed and clearly outclassed, and thousands fell victim to superior American military technology, but they did put up a credible fight on occasion. They were not to be underestimated. As is the case today, unwary, naive, and presumptuous American commanders periodically learned that lesson the hard way.

During the week of 23 Sept 1901, a company from the American Ninth Infantry, under the command of Captain Thomas Connell, had landed on the small island of Samar and moved into the seaside hamlet of Balangiga in an effort to rid it and the surrounding countryside of guerrilla fighters under the leadership of Filipino General Lukban. Just as the unit had settled in and started operations, news reached Captain Connell that President William McKinley had been assassinated. McKinley had been shot in the stomach on 6 Sept by Leon Czolgosz while visiting a trade show. He died of a systemic infection eight days later, on the morning of 14 Sept 1901, only months after beginning his second term as president. Unexpectedly propelled into the Office of President was a young and impetuous Theodore Roosevelt! Czolgosz was quickly convicted of murder and electrocuted.

Captain Connell decided that the late President McKinley needed to be grandly eulogized by himself during a special Sunday service hastily organized to mark the sad event. Security and all other operations in Balangiga suddenly became relatively unimportant as he labored over the text of his speech.

On Saturday evening, the few American sentries who were on duty noticed many women scurrying about the town's only church. They were all wearing scarfs and heavy over-garments and were carrying small coffins. When sentries asked locals what was going on, they were told that the coffins contained the bodies of children who had succumbed to a cholera epidemic and that a service was scheduled for the next day (Sunday).

Cholera epidemic? The sentries were unaware of one, but they decided not to disturb despondent Captain Connell with the news. Besides, the staunchly religious Captain had instructed them to have nothing to do with local women, so none of the "women" were ever questioned, nor searched.

Sunday morning came uneventfully, and American soldiers were solemnly assembled in the local canteen to mourn the President's passing and listen to Connell's lengthy and painstakingly prepared eulogy. Connell had ordered his men to show up unarmed, as he thought the presence of rifles and pistols would "disturb the solemn atmosphere of the occasion." Weapons were all unloaded and locked up in an armory. Some prudent soldiers disobeyed this ridiculous order and carried concealed pistols with them.

General Lukban had studied the movements of the American Garrison carefully during the previous week, and he knew enough about American culture to know that a religious service would take place on Sunday morning and that most American troopers would be required to attend. The strange "women" in town were actually men disguised as women, and the coffins they carried contained knives and machetes which had been distributed among the men of the village during the night.

As Connell was about to emerge from his hut and deliver his speech, church bells suddenly rang. It was a prearranged signal for guerrilla fighters to attack American soldiers concentrated in the canteen. Many astonished (and unarmed) Americans were hacked and stabbed to death immediately, but many fought back, several drawing their concealed pistols and shooting as many of the attackers as they could. Others fought back with chairs, table legs, even hot coffee! Captain Connell was overwhelmed in his hut. He was murdered instantly, his head subsequently chopped off, and his over-embellished eulogy forever undelivered!

Even though all the officers and half the enlisted men had been killed within the first minutes, a heroic Sergeant Benton took command and ordered a fighting retreat to the shore. Survivors ultimately escaped in boats to the American garrison at Basey. Of eighty Americans, only six escaped unhurt. Another twenty-three were wounded. All the rest were murdered. A rescue force was sent from Basey the next day. The bodies of fallen American soldiers at Balangiga had been stripped, mutilated, and left laying where they fell.

An enraged American General Jacob Smith ("Howling Jake") ordered the annihilation of everyone in the village and the entire island. His orders were largely carried out. He was subsequently court-martialed, but, when he returned to the United Stated, he was hailed as a national hero and defended by many who had served in the Philippine campaign.

Balangiga was the worse massacre of US troops since the Little Big Horn.

Lessons: When outclassed by superior military technology, cunning guerrilla fighters will develop clever ways to minimize technological advantage and still inflict casualties. To expect otherwise is to be naive in the extreme.

There is no substitute for alertness and the constant expectation of pernicious threats. Any time you hear, "Relax! There are no threats to your safety here," don't believe it. Let your guard down at your peril. Don't depend upon others for your safety!

Never give up. Never give in. As Sergeant Benton demonstrated, even the most desperate of circumstances can be salvaged when aggressive leaders step up to the plate and lead the way.

To be unarmed is to be defenseless. We teach students to carry discreetly concealed pistols in an effort to hide the fact from casual observation of the general public. My friends overseas right now tell me they have to carry concealed in order to hide the fact from their own commanders!

We see from the foregoing that even that tactic is nothing new. God bless them!

29. Other

28 Nov 2002

"Victory Disease," Overcoming heavy odds at Blood River (Dutch-Zulu War, 1838) and Majuba Hill (First Anglo-Boer War, 1881)

In the Eighteenth Century, Dutch settlers on the South African Cape, like so many other European colonists, swiftly developed a sincere dislike for the British. The British had an annoying habit of making second-class citizens out of all non-British. The British, of course, knew local Dutch farmers lacked the political sophistication to develop the area into a legitimate economic entity. The British were also interested in displacing German influence in the area, so they were not about to pack up and leave. The indigenous Khoisan (Bushmen) were scattered and disorganized. Unlike the ferocious and cunning Iroquois and other Indians in North America, they provided little effective resistance to European settlers. Even the fierce (and well organized) Bantu, coming down from the North and West, were unable to overcome the overwhelming European advantages of the mounted infantryman, the rifle, and the "lager" (a circle of wagons).

Dutch settlers called themselves "Boers." The term translates to "farmer." The British used the term also, but to them it meant "pig."

Many Dutch left the Cape for the South African interior. Others pushed eastward up the coast. All wanted to get out of the reach of British influence. In fact, the "Great Treck," starting in 1836, took on a profound religious significance with the Dutch. They were convinced that the British had been sent by the devil to harass them (British soldiers did wear red uniforms, after all), as they regarded themselves as having been favored by God.

Their beliefs were bolstered at the Battle of Blood River in December of

1838, where Dutch "Voortrekkers" decisively defeated a large Zulu (Bantu) army. Their commander, Andrius Pretorius, cleverly set up a defensive lager in the elbow of the Blood River, frustrating the classic Zulu double envelopment. The tiny Dutch contingent could have easily been wiped out but was instead victorious. They all took this as a divine sign that their presence in the interior of South Africa was sanctioned from on high.

Between the end of the Napoleonic Wars and the beginning of WWI, there were only seven years when British soldiers were not actively engaged in a foreign war. British soldiers have always been excellent fighters. As with the Romans before them, their ability to function as a coordinated team was matched in few other armies. However, British soldiers have never been excellent, nor even respectable, riflemen. Individual marksmanship among British infantrymen has always been poor. Individual marksmanship was never considered important to British commanders. To them, a military unit that is so cohesive that it moves and functions as a single entity was far more important than individual competence with weapons.

In stark contrast, the "army" put together by Dutch farmers in South Africa couldn't have been more different. Dutch religious views made it impossible to have an officer corps, as each man considered himself the equal of every other. So, officers and NCOs were elected and only served in their position so long as their "subordinates" kept them there. Obeying orders was optional, as each man served as a volunteer. He could leave and go home any time he wanted. No one was paid. Thus, large, coordinated military movements, such as an infantry, bayonet charge, were unlikely, because commanders knew many of their men would consider the order stupid and refuse to take part. There was no infantry/calvary coordination, as all Boers reported for duty mounted on their own horses. Their "army" was all cavalry! There were no uniforms either. Except for a cartridge belt across their chest, they all looked like civilians. It all sounded pretty disorganized, but, under the Boer system, dithering buffoons did not stay in positions of command for long, and there was no mercenary attitude. Everyone was there fighting for his home and way of life. As hunters and frontiersmen, Boers were masters of camouflage and stalking. They were hard to locate and harder yet to fix in position.

The British considered the whole thing laughable. Any such disorganized mob would be quickly brushed aside by a professional army. They were in for a surprise, and the First Anglo-Boer War would signal the world that, as would be repeated during the Russo-Japanese War (1904), a small, but agile and highly motivated military force, armed and trained with modern weapons, could handily defeat a much larger force that was unwieldy, outdated, overextended, armed with obsolete weapons, indecisively lead, and most of all, arrogantly overconfident.

Because Boers were all mounted, they moved quickly. They lived off the land, so they didn't require long, vulnerable supply lines. They had the irritating habit of suddenly appearing where no one expected them and just as suddenly melting away as if they had never been there. Their ability to move without being detected was uncanny and was the subject of many exasperated comments by British commanders.

Most importantly, each and every Boer was a suburb rifleman. Having subsisted by hunting native planes game all their lives, poor marksmanship would not be found with any of them. Rifles were not issued. Every man brought his own, and they were never shared. To a Boer, his rifle was his most important personal possession, and it was always well maintained, close at hand, and kept in a high state of readiness. Rifles stayed with their owners at all times. They were never gathered up in stacks or locked away in buildings, as was common procedure with the British.

Their ability to hit, and hit consistently, at extended ranges was something the British had never had to deal with before. British infantrymen were no match for them. Boer riflemen could hit effortlessly at ranges where a British soldier would never even attempt a shot. And, Boers were masters of cover and concealment. They would always fire from expertly camouflaged positions. Precious little was ever exposed. Thus, volley after volley thrown back at them by ranks of British soldiers had scant effect.

Boers inherently disliked all governmental trappings characteristic of Western Civilizations (for which the British were so well known), like multilayered bureaucracies, committees, and endless meetings. They made decisions quickly, with a minimum of discussion. Accordingly, when superior military equipment became available, they didn't dither, they acquired it as fast as they could. A company in England, Westley-Richards, began making a revolutionary breech-loading rifle beginning in the 1860s. It was an improvement of the American Peabody Rifle and, in turn, influenced the British Martini-Henry Rifle, which was eventually adopted by the British (as always, too late for it to make any difference).

With the Westley-Richards breechloader, Boer riflemen could dramatically increase their rate of fire over the that of muzzle loaders that the new rifle replaced. This capability made them essentially invulnerable to bayonet and cavalry charges which formed the very foundation of British tactical doctrine. In addition, metallic cartridges were not as susceptible to environmental deterioration as was loose gunpowder, caps, and balls. There was nothing not to like, so, without hesitation, Boers acquired good quantities of Westley-Richards Rifles immediately upon them becoming available and used them with devastating effect during the First Anglo-Boer War, which began in 1880 and lasted less than a year. By the time the British followed suit (with the Martini-Henry), Boers were acquiring the new, and vastly superior Mauser rifle, just in time for the Second Boer War, which began in 1899!

In 1852, the British, recognizing the potential problems associated with armed conflict with Dutch settlers who had fled to the South African interior, extended autonomy to Boers who had settled beyond the Vaal River (the "Transvaal"). Autonomy was also given to those settled between the Orange and Vaal Rivers, who promptly claimed the title of "The Orange Free State." In the years following, neither "state" ever emerged significantly from anarchy. When they couldn't pay their bills, Britain reannexed them. Both were promptly invaded by high-handed British bureaucrats. Predictably, by 1880, anti-British riots were out of hand, and Boer farmers organized into mobile military units (called "commandos"), began seizing British outposts by force.

Things came to a head when a British infantry column, dispatched late in December of 1880 to reinforce the garrison at Pretoria, was ambushed at a river crossing called Bronkhorst Spruit. A single Boer messenger, under a white flag, commanded to column to turn back. The column dithered. A crescendo of gunfire suddenly erupted from carefully concealed positions on the column's flank. British troopers fell like dominos. One hundred and twenty of them were hit within two minutes. Few ever saw the enemy. Fewer still even got off a single shot. Most died at the scene, including the British commander, Colonel Anstruther. The Boers, with their new Westley-Richards breech loaders, lost only two. The First Anglo-Boer War had well and truly begun!

Back in the Cape Colony, British General George Colley was informed of the Bronkhorst disaster on Christmas Day. He was furious! He was also fearful that peace would break out before he could have his revenge on the brazen and impudent Dutch. Without delay, he organized a regimental-sized expeditionary force and started north to confront the Boers directly. He found them sooner than he expected! At a narrow pass in the Drakensburg Mountains called Laing's Nek, he confronted a large Boer commando, led by Peit Joubert. Apparently forgetting that the enemy was now armed with rapid-fire, breech-loading rifles, Colley assaulted them with a volley, followed by a bayonet charge. The charge never got close to any of the Boer trenches. Once again, British troopers were gunned down wholesale. Colley lost over three hundred men within minutes and was forced to withdraw. A short time later, leading a new force, Colley met disaster afresh at a plateau called Schuin's Hoogte, where, once again, accurate Boer rifle fire from covered positions decimated both his infantry and artillery crews. Again, he was forced to withdraw, this time abandoning both his wounded and his big guns.

Colley was now even more determined to defeat the Boers, in order now to avenge the triple disasters at Bronkhorse and more recently at Laing's Nek and Schuin's Hoogte, where he had been personally in command. In February of 1881, Colley led six hundred men up Majuba Hill in the dead of night. Majuba Hill commanded a vast area that Colley believed he could dominate with artillery. This time, he took the Boers by surprise. He was firmly in position on the summit before any Boer realized it. Eighty Boers volunteered to assault the hill and dislodge the British force, which was four times their number! They had no artillery, and the slope was too steep for horses. No one would have given them much of a chance of success, but, fearless, they went forward anyway.

Majuba Hill is actually a small, volcanic mountain with a crater at the top. On the rim of the crater, are three high points, all of which the British occupied. Many British soldiers were Gordon's Highlanders, fresh from fighting in Afghanistan. Imagine their surprise when several dozen Boers suddenly stood up and fired at them from just outside their perimeter. The Boer commando had snuck into range virtually undetected. Forty Scots dropped dead nearly at the same instant. The rest wavered, then ran! Many were picked off as they scurried toward the hill occupied by Colley and his staff. Colley was advised to make a bayonet charge, but, remembering Laing's Nek, decided to wait until the Boers charged. He would then give them a volley and overrun them.

The Boers never charged. They never even stood up. They just maneuvered and fired from covered positions. British soldiers continued to fall. Return fire was ineffective. Colley himself finally stood up to rally his troops. Instantly, a bullet struck him in the head, and he fell, lifeless, to the ground. Over two hundred British troops were killed that day. Nearly all were wounded. The Boers suffered one killed and five wounded! The British army was in shock. No one dreamed such a thing was possible!

As a result of Colley's quadruple disaster, the Transvaal was again granted independence, but the agreement was unworkable, and a second Anglo-Boer War would break out eighteen years later.

This time, the Boer's luck would run out.

Lesson: As we say are the poker table, "Don't mistake good cards for brains." Unlikely victories, particularly in combination, can cause some to think they can't lose, or that God won't allow them to lose. It's called "victory disease," and it is nearly always fatal. Seasoned warriors know the bad always comes with the good. In poker, winners are not the ones who get dealt good hands. They are the ones who play the poor hand well.

6 Jan 2003

The Age of Kings winds down:

The War of Spanish Succession, 1701-1713

Louis XIV claims the longest reign of any sovereign in Western Europe, before or since. He was King of France for sixty-two years, from 1653 to1715. Like most kings of the time, Louis was conceited and vane, but he was a skilled and durable monarch. So skilled, in fact, was Louis at psychologically outmaneuvering his opponents, that Latin (which had been the official language of European diplomacy since Roman times) was gradually replaced with French.

In no small measure due to the arrival of gunpowder, Seventeenth- and Eighteenth-Century Europe's national armies were becoming "professional," unlike the loosely organized rabble of previous centuries. Armies could no longer "live off the land." They had to be expensively equipped, trained and drilled, and, most importantly, continuously supplied. Suddenly, strategically located supply bases (called "magazines") and protected supply lines became critical to the success of any military operation. On the defensive side of the equation, cutting the enemy's supply lines and assaulting his fortified magazines became military imperatives. Accordingly, siege tactics underwent a great deal of refinement. French military genius, Sieur de Vaubanbecame renowned for his innovative siege tactics. Vauban's influence on both siege tactics and the design of fortifications spanned the centuries, right into modern times, and can be seen at Dien Bien Phu and even at Khe Sahn.

Delusions of grandeur and of catastrophe were common among sovereigns of the period. Kings looked upon warfare as a chess game. Through invasions, open battles, sieges, and peace conferences, they would casually dabble in military adventurism as a way of demonstrating their intellectual superiority or the fact that they personally were favored by God (mostly the latter). In fact, when military fortunes went badly, kings were often convinced that God was punishing them personally for some latent character flaw. Organized religion was also a factor. England and Holland were mostly Protestant. France and Spain were Catholic. In the minds of kings, the issue of which was the "right" religion had to be settled by force. As a result, untold suffering was endured by religious minorities in all four countries, particularly in contested regions.

During Louis' reign, principal players in Europe included England, France, Holland, Spain, Austria, and Sweden. At the end of his reign, only France and England would remain, and their mutual animosity would spill over into the New World. Germany and Austria, despite the best efforts of the Habsburgs, would remain fragmented until the Nineteenth Century (Louis' recurring nightmare of a united Germany would have to wait another two hundred years). As their adeptness at allying with each other and with major players declined, the fortunes of Holland, Spain, and Sweden languished. None would ever occupy center stage again.

In the Seventeenth and Eighteenth Centuries, when an elderly king was in poor health, and the forthcoming vacant throne was hotly contested among various claimants, all insisting that they were the legitimate successor, armed conflict was likely before the matter was settled. Called "dynastic wars" or "wars of succession," these conflicts garnered the interest of neighbors who often had a self-serving interest in the outcome. Further complicating the matter was the fact that all of Europe's royal families were interrelated. Every sitting king had uncles, brothers, and cousins in the courts of other nations. Most were married to princesses from other royal families. Political intrigue and posturing went with the territory.

Spain (by 1700 little more than the bankrupt shell of a once-great empire) under tottering King Carlos II was thus of great interest to France,

England, and Austria. Carlos had no sons. The son of Louis XIV of France and Archduke Charles of Austria were both active contenders for the Spanish throne. The prospect of a Spanish/French alliance alarmed the English, but they were equally unenthusiastic about the Spanish and Austrian branches of the Hapsburg family reuniting. In his last will, Carlos, in an attempt to placate his neighbors, decided to name one of Louis' grandsons (the teenage Duke of Anjou), as the next king of Spain (the Duke of Anjou did, in fact, become the next Spanish king, King Philip V). Carlos reasoned that The Duke of Anjou, not being in line for the French throne anyway, would be an acceptable choice to all his neighbors. He was wrong! The English and Dutch were aghast. Austria, with their native son snubbed, declared war!

During what would be known to history as the "War Of Spanish Succession," John Churchill (The Duke of Marlborough) and Prince Eugene of Savoy made a formidable team. After a string of impressive victories against Louis, they combined once more in their siege of the French fortified magazine at Malplaquet in late summer of 1709. Seeing this as an opportunity for a decisive, war-ending victory, Marlborough and Eugene held nothing back, and the English/Dutch alliance won the day, but at great cost. In fact, Wednesday, 11 Sept 1709 would go down in history as the bloodiest day of fighting in Europe since the invention of gunpowder. Ten thousand French casualties. Three times that many allied troops. Allied losses were so great that Marlborough's plans to subsequently march on to Paris had to be scrapped. The French commander, Villars, wrote to Louis, "If it pleases God to give your enemies another such 'victory,' they will be ruined." After receiving the news from Malplaquet, Louis said (in a typical display of personal vanity), "I am infinitely miserable."

After Malplaquet, Marlborough lost his nerve and avoided all contact with the French. A change in the mood of fickle Queen Anne in England simultaneously landed him on the wrong side of the political fence. He was eventually relieved of command. In fact, following Malplaquet the

English/Dutch alliance disintegrated, after which Holland slipped into permanent irrelevance. With George I replacing Queen Ann in 1714, Marlborough's stock was restored, and he was given command of the army once more, but he never dabbled in politics again.

The war officially ended with the Peace of Utrecht in 1713. In the American colonies, however, hostilities between French traders in Canada (along with their Indian allies in America) and English colonists, also in America, continued. Called "Queen Anne's War," Anglo/French and Anglo/Indian bitterness it created continued to smolder until it erupted anew during the French and Indian War forty years later.

Louis XIV died at the age of seventy-seven in 1715. France would see only two more kings, and the French Revolution would then sweep the last from the throne forever. When the revolution burnt itself out and Napoleon came to power, he never called himself a king.

On his deathbed, Louis gave this advice to great grandson (in line for the throne), "Try to remain at peace with your neighbors. I have loved war too much."

Lessons: Entertaining the thought that God is directly intervening in your life by bringing you good fortune or bad is delusional. Delusional thinking is flirting with mental illness and must always be quickly dismissed. This world operates on a cause-and-effect system and orienting one's thinking logically in that direction is a prerequisite for any kind of success.

During his life, everyone will experience good fortune and bad. With God, it is nothing personal! Persistence and the determination to fight through it all is the best personal philosophy. Even then, there are no guarantees!

As Abraham Lincoln put it:

"If I were to attempt to even acknowledge, much less respond to, every denouncement of me, this Shop might as well be closed to all other business.

I do the best I know how, the best I can. When I am incompetent, I deserve to be fired. Until then, I intend to keep going forward until the end.

When the end brings me out all right, what has been said against me, along with those who said it, will be quickly forgotten and amount to nothing.

And, when the end brings me out wrong, I will be profoundly rebuked throughout history, and ten angels descending from Heaven swearing I was right will make no difference"

5 Mar 2003

Crusades, then and now.

By the end of the first Millennium, the Great Barbarian Invasions of the last five hundred years were ending in Western Europe. The wild Huns had now become Christians and had settled in Hungary, abandoning their nomadic ways, as had the Gauls, Goths, etc. Western Europe was settling down, setting the stage for technological advancement that had been impossible during continuous displacement and conflict.

On Wednesday, 27 Nov 1095 Pope Urban II made an announcement and challenge in Claremont, France. He said to those gathered to hear him, "... either renounce your knighthood now, or go forward boldly as Knights of Christ and rescue the Eastern Church." The enthusiastic response was, "God wills it!" The era of the "Crusades" (loosely translated, "Cross Wars") had begun.

In 330AD, Roman Emperor Constantine moved the capital of the Empire from Rome to Byzantium. Not surprisingly, the city was renamed "Constantinople" after him, and also not surprisingly, the Empire split into Eastern and Western Segments in 395AD. The Western portion lasted only until 487AD when the City of Rome finally fell to invading barbarians, and imperial succession came to an end, but the Eastern Roman Empire continued on for many centuries until Ottomans (Turks) conquered Constantinople in 1453 (renaming it "Istanbul"). While it lasted, The Eastern Roman Empire became known as the Byzantine Empire, after Byzantium, the Greek name for a city on the Bosporus (a strait connecting the Black and the Mediterranean Seas). Byzantines were well and truly

Romans and Christians.

**

The Ottomans (from Osman, the first sultan) were nomadic tribes who (like the Huns before them) migrated to the Middle East from central Asia. Ottomans had interbred with Huns to create "Turks." Turks then assumed the role of the military arm of the Islamic religion. When Byzantine Emperor Romanus Diogenes was ignominiously defeated (and captured) by the Turks at the Battle of Manzikert in 1071, his nervous successor, Alexius Comnenus, pleaded with Pope Urban II to rescue the Eastern Empire from the Turkish invaders, and, in the same, breath, offered to reunite the Eastern and Western branches of the Christian Church. The offer was too tempting for Urban to resist. Sending bored barons and knights from Western Europe to the Mideast would put an end to their incessant local, territorial squabbles. It would also have a uniting effect on all of Western/Christian Europe, vastly increasing the power and influence of the Papacy.

The City of Jerusalem had been under the control of Arabs since the Islamic movement had begun in the Seventh Century. However, Christians, Jews, and Islamics all continued to occupy the area and, to a degree, tolerated each other. They had a lot in common. They all venerated the Old Testament and the prophets and progenitors named therein. Christians and Jews just failed to recognize the last great prophet, Muhammad. Christian pilgrims from Europe came to Jerusalem regularly, and, like tourists everywhere, were welcomed and treated well. However, Turks had zero tolerance for anyone outside the Islamic religion, and that was the problem. Their influence gradually made life for local Christians and Jews alike disagreeable and dangerous.

The first rescue effort (called the "Vanguard" or the "Peasants' Crusade"), poorly organized and more intent in murdering local Jews than saving Jerusalem, arrived in Constantinople in 1096. Anxious to get rid of such an unruly mob, Comnenus delivered them to the Turks who, with their horse-mounted archers, effortlessly wiped them out. As a result, for hundreds of years afterward, the term "Byzantine" was used to describe a devious and untrustworthy ally.

The next year, a more orderly and prepared army arrived and immediately set out, capturing Antioch and ultimately Jerusalem itself (massacring most of its residents in the process). Called the "Barons' Crusade" or the "First Crusade" (by now, everyone wanted to forget the Peasants' Crusade), this army featured the vaunted French heavy (armored) cavalry as well as the crossbow. An enduring Christian presence was subsequently established in Jerusalem and the surrounding area. The daunting task of defending these Christian outposts would be the objective of subsequent crusades. All would ultimately fail, and, by 1300, Christian influence in the area ended permanently.

Arrows launched by horse-mounted archers had scant effect on armored French knights and their armored horses. But, bolts launched from crossbows outdistanced arrows and easily penetrated light armor used by Turks. Turks were also plagued by internal squabbles. In Antioch, the spear, said to be the one used to impale Jesus, was found and served as an effective rallying symbol for the crusaders. In fact, the "Battle of the Lance" routed Turks defending the city.

Crusaders received their decisive defeat in 1187 during the Battle of the Horns of Hattin at the hands of Salah ad-din Yusuf ibn-Ayyub. His people knew him as Al Nasir ("The Victorious"). Crusaders knew him simply as "Saladin." Saladin was devoted to the Islamic faith and absolutely opposed to Christianity in any form. He was also a master strategist. At Hattin, Crusaders, out of water and suffocating within their armor, were first disconnected and then overwhelmed by Saladin's mounted archers.

Saladin then went on to retake Jerusalem and ultimately dried up all Christian influence in the area. Crusades would go on for several more centuries, but Saladin's victory at Hattin ultimately sealed the fate of the Holy Land.

Unhappily, Saladin sealed his own fate and that of his people in the process. He and his successors in the Islamic world suffered from "victory disease." They continue to suffer from it to this day. Their abject hatred of Western/Christian Civilization caused them to unequivocally reject every Western influence. They fell hopelessly in love with their victorious horse-mounted archers, so much so that they resisted all advancements in military science from that point forward.

Following Hattin, Islamic control of land routs in the Mideast forced Western European powers, particularly England and Portugal, to harness wind power and develop a blue-water merchant fleet and navy. Rejecting all this, Arabs and Turks continued to use galleys and slave rowers, which left them out of the New World. Islamics also rejected gun powder and the new weapons it made possible. They became frozen in time, steadily falling behind the technological advancements of the Western World.

They received their first hint on Tuesday, 2 Feb 1509. The Sultan of Egypt dispatched a fleet of galleys to confront Portuguese ships that were sailing around the African Cape and plying the waters of India. Egyptian galleys made contact with Portuguese ships off the tiny island of Diu in the Gulf of Khambhat. Islamic sailors and marines aboard their galleys marveled at the tall-masted sailing ships, but they pressed their attack anyway. Before they got anywhere near the Portuguese, ship-mounted cannon blasted the Egyptian galleys to splinters! As a result of that and several other, similar incidents, the Egyptians were in shock. They finally, and reluctantly, realized that they had allowed themselves to fall hopelessly behind Western/Christian civilization.

Unfortunately, the response of the Islamic world was, and continues to be, more hatred, rejection, and isolation. This attitude has blunted the potential and aspirations of the descendants of Saladin and, in our time, brought about a new confrontation between the Islamic world and Western Civilization.

Lesson: Never be too sure you're right. Rejecting useful technology on religious/ideological grounds will always lead to ossification, isolation, and ruin. We must continuously face to the front and never allow ourselves to think that we are perfect and that improvement is thus impossible.

25 Mar 2003

The Fourth Crusade and the Fall of Constantinople, June, 1205AD.

Crusades from Western Europe to the Mideast continued throughout the 1200s, including a pathetic "Children's Crusade," (1212) where thousands of children and gullible teenagers, convinced that the Mediterranean Sea would miraculously part and let them march through it, starved or froze to death before they ever reached the shores of the Mediterranean.

The Fourth Crusade (1202-1205) resulted from the disastrous failure of the Third Crusade to recapture Jerusalem (which had been captured and consolidated by the first Crusade then reconquered by Moslems). Once again, the original intent of the Fourth Crusade was to proceed with a vengeance directly to Jerusalem and the Holy Land.

But global politics got in the way:

By 1200AD, both Saladin (always in poor health) and his nemesis, Richard ("the Lionhearted"), had both died. Their truce had expired, and Moslems had resumed their (re)conquest of Western Christian vestiges in the Holy Land, as noted above. In France, predictably, a new Crusade was proposed, but its leaders decided to attack Moslem Egypt first instead of proceeding directly to Jerusalem. Egypt was fat, slothful, and its military was considered impotent. Much easy booty laid there for the taking. Afterward, of course, off to Jerusalem they would all go!

The autonomous port city of Venice on the Italian Peninsula was selected as the launch point. Crusaders would march from France to Venice, then set sail for Egypt. However, grand political intrigue developed between Enrico Dandolo, the elderly Doge (Duke) of Venice, Pope Innocent III, and the Crusaders themselves. Dandolo saw Venice's influence expanding throughout Crete, Rhodes, and the entire Aegean Islands. Innocent III saw the Eastern Church finally coming under the influence of the Western Church. And naive Crusaders saw Jerusalem recaptured and themselves getting rich.

Egypt was a strong trading partner with Venice, and Dandolo did not want to antagonize them, so he told the Egyptians not to worry about Crusaders (which he regarded as just another dreary invasion of barbarians), as he planned to divert them. As usual, Crusaders had arrived depleted in number and broke. In exchange for a fleet of ships and transport to Egypt, Dandolo asked the Crusaders to attack the Hungarian City of Zara. They agreed and besieged Zara, which surrendered and was quickly annexed by Venice.

Unfortunately, Zara was a Christian city. Upon hearing of the attack, Innocent III threatened to excommunicate all Crusaders for attacking other Christians. However, excommunication might be suspended if Crusaders would agree to go to the city of Constantinople, throw out the existing Byzantine rulers, and install a new ruler who would be aligned with the Western Roman Church. Of course, Byzantines were also Christian (albeit Eastern Christian), but Innocent thought that, after all, ultimate Christian consolidation was worth spilling a little Christian blood.

Gullible and confused, Crusaders allowed themselves to be thus used. On 20 Apr 1203 (Easter Sunday), the Fourth Crusade sailed off to Constantinople. Since the ill-fated "Peasants' Crusade" had been double-crossed by Comnenus in Constantinople in 1096, the term "Byzantine" had been used to describe a devious and untrustworthy ally. These new Crusaders were anxious.

The expedition's chronicler, Geoffroy de Villehardouin, described Constantinople:

"I can assure you that all those who had never seen Constantinople before gazed intently at the city, having never imagined there could be so fine a place in all the world! There was indeed no man so brave and daring that his flesh did not shudder at the sight."

Sitting on the dividing line between East and West, Constantinople's citizens had always considered themselves to be the Eastern

Roman/Christian outpost. Since 330AD, Constantinople had been besieged by Goths, Huns, Slavs, Magyars, Arabs, and Russians. To their sorrow, all had found its famous double walls (on the land side) impregnable. In addition, across the harbor Byzantines had stretched a great chain.

The Crusader's initial show of strength caused the city's defenders to draw back. Crusader's ships continually rammed the chain until it finally broke. A fierce struggle subsequently led to the capture of the city. Crusaders installed their new sovereign, but relations quickly soured, and the Crusaders were eventually driven from the city. They then attacked anew, the second attack of Constantinople within two years. This time, just as the attack was failing, a serendipitous gust of wind caused two assault ships to make contact with the city's (single) defensive wall on the sea side. The wall was breached, and the attack was ultimately successful.

A new Roman-Church-friendly governor was installed in Constantinople. Donaldo was happy, as the Greek Islands were now under his influence. Innocent III was happy, because the Western and Eastern Churches were finally reunited. The Crusaders were happy, because they had a great victory to their credit (as might be expected, the reconquest of Jerusalem was, once more, shelved indefinitely).

However, happiness was short lived. In 1261, Byzantines recaptured Constantinople, permanently ending Latin occupation. Eastern and Western branches of the Christian Church split once more, never to rejoin again. By 1300, the entire Holy Land, including Jerusalem, was back in Moslem hands. In 1453, Moslem Turks captured Constantinople, renaming it "Istanbul." By 1500, Western Europeans were worried more about keeping Moslems out of Europe than they were about military adventures into the Holy Land.

The "Age of Crusades" had ended!

The big loser was the Roman Catholic Church. It had been lured away from its mission, abandoned its moral authority, and was seduced by wealth, power, and influence- all the worldly effects that seduce the rest of us. From the Fourth Crusade onward, the Roman Church worried more about the size of its armies and bank accounts than it ever had about the number of souls it saved. Like the Moslems it purported to despise, it henceforth enforced its will through violence instead of principled persuasion. The Roman Church fell (only too willingly) from its celestial throne, setting the stage for Martin Luther and the Reformation of 1520, disastrous religious wars that followed and immolated most of Western Europe during the 1500s and 1600s, and ultimately the end of the Holy Roman Empire.

Lesson: It's easy to get sidetracked. All need a mission. Without one, we drift aimlessly. Treat it casually at your peril!

13 May 03

Eighty-seven years before the current Gulf War: British Soldiers in Basra and Kut, April 1916

In 1915, there was no nation of Iraq. Iraq did not become a sovereign nation until 1932. All of Mesopotamia was then part of Turkey, and its inhabitants were all "Turks." Turkey had allied itself with Germany after the Great War broke out, and England suddenly had a direct interest in oil and oil pipelines in what was now enemy territory. All the real action was, of course, in France and other parts of Western Europe, but military backwater campaigns, such as this one in Mesopotamia and the ill-fated and simultaneous Gallipoli Invasion (also against the Turks) were seen as critical to the success of overall war effort. Unfortunately, as with most military sideshows, both campaigns proved disastrous debacles.

Sir Charles Townshend had gained minor notoriety in 1895 when his command held out for two months against native tribesman at Chitral on the western Indian frontier. Despite his personal ambition (noted by all who knew him), his reputation still did not suffice to get him assigned to a line unit in the main battle area in Europe with the onset of the WWI. In fact, his disappointment at being shunted off to a military backwater like

Mesopotamia was the subject of many bitter conversations with contemporaries.

However, trying to make the best of a disappointing assignment, Townshend, with his British 17th Infantry Brigade, charged northward from Basra, up the Tigris River. Turkish resistance was weak and disorganized. Turkish units mostly retreated without firing a shot. Within a month, Townshend was already halfway to his goal, Baghdad, with only light casualties. In mid-1915 British victories elsewhere were scarce, so, when the news reached England, Townshend became an instant hero.

Noting Townshend's success, the German Kaiser sent Field Marshall Colmar van der Goltz to organize the Turkish resistance and defend Baghdad. Van der Goltz was no amateur! Immediately, and for the first time, Townshend encountered effective resistance at Kut as he resumed his march north. He also started taking significant casualties. Kut was finally taken, but at a heavy price.

Townshend's superior, General Nixon (back in Basra) apparently jealous of Townshend's success, did little to support the advance. He obviously did not want to see the arrogant Townshend promoted. Fresh troops and supplies were thus not moving north.

Townshend continued pushing north anyway, as far as the ancient Persian battlefield of Ctesiphon, affectionately called "pissed upon" by his troops. Ctesiphon (twenty miles southeast of Baghdad, called "Al-Madain" today and largely a ghost town, then and now) had marked the southern limit of the ancient Roman Empire (under General Belisarius) many centuries before. Ctesiphon, as fate would have it, was also to mark the northern extreme of the British advance. Turkish resistance was now so heavy, that for Townshend to continue on to Baghdad without substantial reinforcements, was out of the question. Townshend had no choice but to fall back.

At this point, Townshend's contemporaries note that he began to suffer a nervous breakdown. With a great "Victory at Baghdad" now a rapidly fading fiction, Townshend's hoped-for "place in history" began fading away too. His overblown personal vanity couldn't handle it. As a result, Townshend began to fall apart mentally.

The smart thing to do would have been to retreat all the way back to Basra, which could be accomplished in relative safety and with comparative ease. Once there, Townshend could consolidate, reinforce and resupply, get competent medical care for his many casualties, and confront Nixon face to face. However, Townshend's faltering ego wouldn't allow such an ignominious retrace of his once-glorious advance. He unwisely decided to dig in at Kut, where he was quickly surrounded by Turkish forces. The possibility of significant resupply and reinforcement was expeditiously eliminated by the Turks. Townshend naively thought he could hold out at Kut until he was relieved by a column coming north from Basra. Salvaging his personal reputation now occupied him completely. Perhaps the newspapers in London would make him a hero after all, comparing him to Gordon at Khartoum.

No such luck! Adopting the tactic of slow strangulation, the Turks brought in German artillery and began what they calculated would be a long siege. Townshend's sanity deteriorated rapidly. He became giddy and whimsical, secluding himself most of the time and issuing conflicting and illogical orders on the rare occasions when he came out. He should have been relieved. He made no effort to break out from Kut, and an overdue relief column from Basra, despite heavy casualties, was unable to break through from the other direction.

On Saturday, 29 April 1916, Townshend surrendered his entire command at Kut to the Turks. Townshend's men subsequently endured a brutal captivity in Turkish POW camps. Most, succumbing to starvation, disease, and murder at the hands of Turks, did not survive to return to England. Townshend himself was meanwhile wined and dined in Constantinople as a guest of the Turkish government (who apparently regarded him an intriguing curiosity), all the while caring nothing for the lot of his men.

Townshend was eventually returned to England where he vainly attempted, for the rest of his life, to justify his conduct. His attempts fell on deaf ears, particularly those of the few survivors of his ill-fated command.

Lesson: Personal vanity has ultimately disgraced the careers and reputations of more than one commander. Custer, Townshend, and Percival all suffered from "delusions of grandeur" so burdensome that they literally lost their minds when they saw it fading away. A sound mind carefully combines humility and audacity. Tilt too far in either direction, and disaster awaits, as we see from the foregoing.

29 Sept 2003

Adrianople, 9 Aug 378AD, "The End of All Humanity, the End of the World"

As the Millennium turned, the stormy Hsiung-nu tribe in China, like so many others, found itself progressively marginalized. By the year "Zero," they were fleeing in the only direction they could- west. "White" was the term Chinese used for westerners, and the Hsiung-nu were so dubbed. Eventually, the "White Hsiung-nu" lost contact with their relatives in China as they fled further west. Ultimately, they ran into the eastern frontier of the Roman Empire. Finding "Hsiung-nu" difficult to pronounce, the Romans shortened it to "Huns." The Huns had no "infantry." All were mounted, and their horsemanship was astonishing. They had no heavy metal armor, so their mounted formations were swift and light, and their horse archers and stirrup-equipped lancers were second to none.

The Huns shared a great strength with the Romans: ingrained military organization. Roman legions had easily defeated fractured tribes in Germanica (Goth), Gaul (France), Iberia (Spain), Africa, etc. So intense was their internal squabbling, Barbarians (the "outer ones") just couldn't unite in any meaningful way to opposed unambiguously ordered Roman Legions. As a result, they were, with a few notable exceptions, "defeated in detail," and all eventually became Roman vassal states. Many of their warriors were subsequently incorporated into Roman Legions.

That eventually became a fatal weakness, in both camps. Both Hunnish and Roman armies gradually became swollen with mercenaries of dubious loyalty. In both camps, what began as an "army of conviction" evolved into an "army of convenience." At the same time, there was an important philosophic difference between the Huns and the Romans: Romans were interested in real estate. They wanted to conquer, colonize, and then form a permeant tax base- in that order. Huns, nomads by nature, were interested only in expanding their warrior hordes. They wanted to conquer, strip the locals of valuables, increase their numbers, and move on- in that order. Neither national philosophy was destined to weather the storms of history.

When the Huns encountered the nimble Alans on the eastern Roman frontier, they defeated them easily. The Huns were organized; the Alans weren't. As they pushed further west, Huns next encountered the Ostrogoths ("Eastern" Germans). They went down to defeat in exactly the same way, as did the Visigoths ("Western" Germans). Many Alan and Gothic warriors subsequently joined the Huns. That was all just fine with Rugila and his nephew, Attila, the Hunnish leaders.

In 330AD, Roman Emperor Constantine, in an effort to ecumenize the Empire, moved the imperial capital from the City of Rome on the Italian Peninsula southeast to Byzantium on the south shore of the Black Sea. Not surprisingly, Byzantium was subsequently renamed "Constantinople" after him, and also not surprisingly, the Empire permanently split into Eastern and Western Segments in 395AD, each with its own emperor, each claiming to be the "legitimate inheritor" of the first Roman emperor, Octavian Augustus', throne. The Western Roman Empire would not survive the following century.

Up until 375AD, Roman Emperors had all been generals, with mud on their shoes, and they thus enjoyed the respect of the Legions. Many, of course, were also world-class lechers, but they were still considered qualified to be emperors by virtue of their personal experiences and accomplishments. That all ended with the succession of Gratian in 375.

Barely more than a teenager, Gratian was neither respected nor feared. The loyalty of the Legions and the unity of the Western Empire simultaneously began to crumble, as citizens looked to religious clerics, instead of civil leaders, for national affirmation. Unfortunately, religious leaders proved themselves no more worthy of respect that had secular ones.

The winter of 406 had been particularly cold, so cold in fact that, on the last day of that year, the Rhine (which was then the eastern border of the Western Roman Empire) froze solid, something that had happened only a few times in recorded history. Goths, Vandals, Alans, Franks, all fleeing the Huns, pored across. Roman legionaries on the opposite shore were unable to stop the horde of "illegal aliens." With its once famous Legions now made up almost entirely of mercenaries, the City of Rome was considered too vulnerable, so the capitol was moved to the compact, walled city of Ravenna. Just in time as it turned out, as Aleric, King of the Visigoths, sacked the City of Rome in 410. With that, the entire Western Roman Empire descended into anarchy. During the year 410 alone, there were six, separate western emperors. Installed via bribes and political intrigue and removed via murder, they monotonously came and went. The record was set by Emperor Sinerich, who, in 415 held the throne for all of seven days, only long enough to murder his rival's children before he himself was murdered.

By 475, the Empire's holdings had shrunk to the Italian Peninsula itself and several regions in southern Gaul that were still "Roman" only because Goths and Huns regarded them as insignificant and hadn't gotten around to invading them yet. The dubious title of "The Last Emperor" was claimed by Romulus Augustulus ("Little Augustus," actually an insult). Ruling briefly from Ravenna, he was knocked off the throne by invading Ostrogoths in 476 (who considered him so impotent, they didn't even bother to murder him). Imperial succession had finally come to a merciful end. In 487AD the surviving vestiges of the once-mighty Western Empire dissolved completely. From that point forward, squabbling Ostrogothic warlords held sway on the Italian Peninsula, but none displayed interest in the (now meaningless) title of "Emperor."

The "Dark Ages" had begun.

The Eastern Roman Empire continued (in name, at least) until Turks conquered Constantinople in 1453, renaming it "Istanbul". Ottomans had interbred with Huns (and others) to create "Turks." Turks then assumed the role of the military arm of the Islamic religion.

It was the Western Roman Empire that Charlemagne (Charles the Great) in Gaul (France) attempted to resurrect with his "Holy Roman Empire" in 800AD.

During the existence of the Roman Empire, there were four notable defeats of the Legions, among many victories:

>An unprepared City of Rome was sacked by Celtic tribesmen in 387BC

>With his famous "amoebic defense," Hannibal deceived and decisively defeated Rome's best at Cannae in 216BC, during the Second Punic War

>Rome's entire 17th, 18th, and 19th Legions were wiped out by Goths at heavily forested Teutoburger Wald in Germanica in 9AD

>Rome's best are again defeated decisively, this time by Huns, at Adrianople on 9 August 378AD (present-day Edirne, Turkey)

The Battle of Adrianople is considered the "beginning of the end" for Western Romans. During the battle, the Emperor Valens himself, was killed. Roman infantry attacked a circle of wagons occupied by Huns (with their Alanic and Gothic mercenaries). Roman formations were then counterattacked by Hun "heavy cavalry" (stirrup-equipped lancers) which left Roman infantry formations in disarray. The defeat convinced Roman war planners, who had for years depended on infantry tactics they had learned from the Greeks, that heavy cavalry had displaced infantry as the "Queen of Battle." As a result, reliance on heavy cavalry would dominate the thinking of war planners throughout Europe for the next thousand years, and, in fact, would come back to haunt Christian Crusaders at Hattin eight hundred years later.

Roman general Aetius returned and defeated Attila and his Huns at Chalons in western Gaul (France) in 451, using his Franks as infantry (the French didn't like horses) and his own Gothic heavy cavalry. However Aetius, fearing another Cannae, made the same mistake that would be made by Union General Meade centuries later at Gettysburg, PA. He didn't finish the fight, and Attila's army was allowed to escape. Aetius would live to regret it, as Attila and his hordes, unpredictable as ever, invaded the Italian Peninsula in 452. Nothing less than a personal conference with Pope Leo persuaded him not to sack the City of Rome.

Attila had an astounding military origination, but only he could make it work. He died (or was murdered) right after marrying a beautiful German woman in 453. Shortly thereafter, his fearsome army began to disintegrate. What remained was crushed at the Battle of Nedao in present-day Hungary (454AD), not by Roman legions, but by Germans. The Hun era in Europe came to an abrupt and permeant end. Most native Huns remained in Europe but never formed a cohesive force again.

Lessons: An army of mercenaries can never substitute for an army of patriots, and sleazy, faint-hearted, and corrupt national "leaders," who do not command respect, can never "stand in" for honorable and fearless heroes. During human history, civilizations come and go, no matter how strong, no matter how advanced. There are no guarantees and no "Divine Protection."

Personal honor, decency, and courage are the basis of any successful civilization. Moral and virtuous citizens for the rock of any successful society. Toleration of, and eventual encouragement of, corruption, iniquity, and sleaziness unfailingly mark the decline of a civilization.

Unfinished battles and unfinished wars will invariably come back to haunt the "victors." Leaving an enemy in tact is a virtual guarantee that you will be compelled to face him again.

We ignore the foregoing at our peril!

28 Nov 2003

A missed opportunity to alter the course or world history, never to present itself again, Nakma Chu, India, 1962

The Himalayas have been disputed territory for centuries. When Communists took over in China at the end of WWII, thousands of Tibetans (we'll never know how many) were displaced and ultimately murdered. To the South, newly independent India's Jawaharlal Nehru became progressively nervous, as Communist China also made territorial claims to large parts of its northern frontier. The warm and fertile plains of India were tempting military objectives and would make a welcome addition to Chinese national dominion, and the Chinese knew it was all ripe for the picking!

With no roads in the area, India's northern frontier posts could only be reinforced and resupplied by helicopter, and the high altitude limited carrying capacity of each sorty to barely more than a thousand pounds. Airdrop was the only other option, and the mountainous terrain and high winds made that a dubious proposition indeed. Significant reinforcements would have to get there by foot from lower altitudes.

Assignment to these posts was considered punishment. Living conditions were wretched. Acclimatization to high altitudes took weeks, so troops were rarely rotated. Pulmonary disorders and temperature casualties regularly reduced manpower to far below published levels. Radio communication with headquarters at lower altitudes was sporadic and unreliable. Maneuvers were rare, as troops had only individual weapons (rifles and pistols). There were no machine guns, mortars, or artillery. Indian troops mostly huddled in shacks and tried to stay warm. By contrast, corresponding Chinese posts were amply supplied via all-weather roads. Hardwire communication was all-inclusive, and manpower was more than adequate. Patrols and other maneuvers were constant and aggressive, and Chinese troops were well equipped with crew-served, as well as individual, weapons and were backed by artillery.

However, China's Mao Zhe-dung, after being surprised by America's unexpected resolve on the Korean Peninsula, was concerned about British (and, by proxy, American) military involvement if he made an overt territorial grab on India's northern edge. On the other hand, he had no fear of a dithering Nehru and his bumbling administration, and, based on the Korean experience, he was pretty sure that Americans lacked the will to use nuclear weapons.

By the early 1960s, Nehru, trying desperately to be taken seriously by the Western Powers, decided to take a stand (of sorts) against provocative Chinese intrusions. Unfortunately, Nehru's "army" was completely outclassed, in every way, by the Chinese. As noted above, India's frontier posts were remote, poorly manned, and miserably equipped. Nehru's pitiable forces had no chance against the well-equipped Chinese, but Nehru, relying on fraudulently optimistic reports manufactured by his generals (who, like generals everywhere, were afraid to tell him the untidy truth), pushed the issue anyway, issuing empty threat after empty threat, hoping to get Mao to back off. The stage was set!

On 10 October 1962, Nehru's general Kaul was personally sent to the area and commanded to "throw out" Chinese troops from disputed territory. Kaul's "strategy" was to have a company of his men march in the open, past Chinese positions, and defiantly "sit" behind them. In an attempt to execute that plan, hapless Indian troopers made it only a few meters forward before being annihilated in less than a minute by an amused, dug-in Chinese battalion. Kaul was boggled! He immediately got aboard the nearest helicopter and deserted the area, muttering that he had to "consult" with higher command. He never returned. On 20 Oct, hordes of Chinese troops, supported by artillery, effortlessly overran one Indian position after another. They were met with virtually no resistance, as thousands of fleeing Indian soldiers were killed and captured.

There was silent panic when the news reached New Delhi, as all the fiction about India's military prowess began to unravel. Newly arriving reinforcements, rushed to the area, were all deposited at low altitude and instructed to climb, on foot, to the area under siege. All quickly succumbed to altitude sickness and the cold. The few that actually arrived had inadequate clothing and only a few rounds of ammunition. It quickly became obvious that there would be no significant reinforcement and thus no credible resistance.

With his new nation on the verge of being swallowed up, Nehru issued a plaintiff call to Britain and the United States to come to his aid. Nehru now knew the dysphoric truth. Without immediate foreign intervention, he could not stop, or even slow down, the Chinese. But, even as intervention was being contemplated, the Chinese suddenly withdrew back to their original positions. The crisis was apparently over.

Mao was advised that the point had been made, and that risking confrontation with Britain and the United States was too big a gamble. It was a colossal and costly miscalculation on Mao's part! Mao had little to fear from either the British or the Americans. By the time British and American forces could be mobilized and transported to India (assuming they could even make a decision to come to Nehru's aid), the invasion would have been over, and China would be in the process of consolidating its new territory. After begging America not to use nuclear weapons in Korea, it would be the height of hypocrisy for Britain to urge their use in India.

In the early 1960s, China could have easily annexed all of India at virtually no cost, citing, correctly, that India had started it. By the grace of God, they missed their chance!

Lesson: Any political leader who believes, without question, what his generals tell him is a fool! Nearly all military disasters are preceded by overly optimistic, indeed fictional, reports, promulgated by generals who are interested only in keeping their jobs. Digesting the lumpy truth is always more difficult than gulping down smooth lies. But, in the end, lies, smooth or otherwise, always unravel. If you rattle the saber often enough, someone will call your bluff. When they do, you better not be bluffing!

13 Dec 2003

Teutoburger Wald (loosely translated, "Battle in the Forest Near the German Village"), Germanica, east of the Rhine, fall of 9AD.

Roman general, Germanicus Caesar, put it this way:

"Iberians (Spaniards) can be impressed by the courtesy of their conquerors; Gauls (French) by his riches; Greeks by his respect for the arts; Jews by his moral integrity; Nubians (Sub-Saharan Africans) by his calm and authoritative bearing.

But, Goths (Germans) are impressed by none of this! They must be struck into the dust; struck down again as they rise; struck again while they lie groaning.

Only so long as their wounds still pain them, will they respect the hand that dealt them!"

Echoed by Heinrich Heine in the Nineteenth Century AD:

"Christianity has occasionally calmed the brutal German lust for battle, but it cannot destroy that savage ecstasy. When once that restraining talisman, the cross, is broken, the old stone gods will leap to life once more."

Germans have "enjoyed" quite a reputation all these years!

After being conquered by Assyrians in 800BC, Babylonians returned the favor, and, in alliance with Medes, invaded the Assyrian capitol, Nineveh, in 600BC and burned it to the ground. In 538BC, Persians, under Cyrus the Great, overthrew Nebuchandnezzar's Babylonian Empire and rushed onto history's center stage. Cyrus subsequently allowed captured Israelites to return to Jerusalem. He even financed the rebuilding of their temple! Cyrus is thus regarded as a great hero by today's Jewish community, even though he was himself never one of them!

Between 492BC and 479BC, Persians, under Darius and Xerxes, tried their level best to stamp out the tiny flame of Western Civilization in Greece, but the valiant Greeks held them off. One hundred and fifty years later, Greeks, under Alexander the Great, turned the tables, conquering all of Persia. When Alexander died (at the age of thirty-two), his vast kingdom was ultimately split up among his squabbling generals.

Under the daring military genius, Judas Maccabeus (called "The Hammerer," he was the author of the famous Maccabian Rebellion, celebrated to this day as Hanukkah), the world witnessed (for the first time since Southern Israel had been conquered by Babylonians in 586BC) an independent Jewish State emerge in Palestine and exist there for seventy-nine years, between 142BC and 63BC. In 63BC, Judea, and all of Palestine, became a vassal state under Imperial Rome, as did all other regions in the vicinity of the Mediterranean. In fact, the entire Mediterranean became a Roman lake! A politically autonomous Jewish state would not reemerge for another 2011 years, when, in 1948AD, the modern State of Israel boldly claimed its independence.

By the year "Zero," the entire Roman world had settled down. Romans kept a tight lid on their expanding Empire, tolerating some political autonomy in vassal states (like Judea), but having no tolerance for open rebellion. Spartacus and his rebel gladiators displayed dazzling success against Roman legions for a time, but even that rebellion was finally put down for good in 71BC. Relative peace would prevail for the next two hundred years, punctuated only by a few famous battles. By far, the most famous was the debacle at Teutoburger Wald in 9AD.

Many tribes and peoples chafed under the Roman yoke, but some chafed more enthusiastically than others! Heretofore, land battles had been disorganized brawls. Romans (learning from the Greeks) brought precise organization to land warfare. So long as Roman Legions could preserve their precise organization and fight on firm ground, they were unbeatable. However, in thick, wet forests or among steep hills, where Roman commanders could not maintain visual contact with their units, they often did poorly. For example, In Britain, the southern portion was well pacified (with the exception of the short-lived Boadicean Rebellion in 60AD), but, in the hilly and rocky north (Scotland), wily Scots proved themselves so troublesome that the Romans simply built a stone wall (Hadrian's Wall), from sea to sea, all across Britain, in an effort to isolate them. Hadrian's wall, built around 100AD, was tangible evidence that there were acknowledged limits to Rome's ability to extend her influence.

Another fractious area was Germanica. At the turn of the Millennium, southern Germanica was pacified west of the Rhine. However, the further north one went, the less "pacified" the territory became. East of the Rhine was the frontier, "barbarian county," and the Rhine River made a convenient dividing line. Germanica was heavily forested and occupied by Cherusci and other Gothic tribes, known for their savagery and brutality, if not their military organization. They were every bit as warlike and cruel as had been the Assyrians (to whom they were probably largely related).

In 7AD, a lawyer, Publius Quintilius Varus, due mostly to the fact that he married a grand niece of Emperor Augustus, found himself appointed governor of this ill-defined area known as Germanica. Varus had a good deal of governing experience, but scant military acumen. He had come north from comfortable and well-pacified Syria with three legions, the 17th, 18th, and 19th under his command. He grossly and foolishly underestimated the danger he and his legions were in.

By fall of 9AD, eager to establish himself, Varus decided to march his legions through contested real estate, east of the Rhine, between the Rivers Ems and Weser. It was his intention to build a fort and set up a winter camp in the area as well as establish protected supply lines back to the Rhine, laying the groundwork for a conspicuous and unmistakable Roman presence. There were reports of sporadic, local disturbances, but Varus was confident he could indiscreetly restore order.

Varus was befriended by Arminius of the Cherusci. Arminius (son of a local nobleman), with his entourage, agreed to serve as guides, scouts, and translators for the expedition. Arminius had been educated in Rome as a captive, was clever, poised, well mannered, spoke fluent Latin, and was intimately familiar with Roman military protocol and tactical doctrine. However, like Dragutin Dimitrijevic, who centuries later would engineer the start of WWI, Arminius was a cunning and duplicitous con man. He nursed a bitter, seething hatred for Romans, but was outwardly congenial and accommodating. While appearing to assist Varus, in reality Arminius had a hidden agenda. He was plotting the destruction of Varus and his entire army. "Arminius" was his adopted Latin name. His family knew him as "Hermann."

Curiously, owing to his Roman upbringing, Arminius' loyalty was never questioned. Varus was completely taken in. He was convinced by Arminius that this expedition would be a cake walk. Accordingly, Varus failed to put his legions on full, or even active, war footing and even allowed officers to bring their families along! Like naive Americans at Pearl Harbor in December of 1941, they were in hostile territory that was on the brink of open warfare, and yet they all acted as if they were on vacation! The complacency of Varus and his staff was inexcusable, and they would all pay a terrible price.

Arminius had engineered an enormous, U-shaped ambush, deep in the thick and swampy forest, deep enough so that a Roman relief column could not arrive in time. He knew an overconfident Varus would not deploy flanking units nor dispatch scouting parties. Attackers could thus get close to Roman columns undetected. He also knew that a noisy and disorganized mobile city of camp followers, consisting of hordes of dependents, servants, caterers, tailors, service people of all kinds, baggage wagons, and prostitutes would follow in the column's train, slowing it to a snail's pace. Amazingly, Varus was urgently warned of the ambush by none other than Arminius' own father-in-law, who despised his underhanded son-in-law for taking his daughter. Incredibly, Varus failed to investigate the claim, casually dismissing it as nothing more than a family squabble.

Deep into the German forest, the going had been uneventful, except for a few harassing attacks by local tribesmen. Then, Varus and his staff suddenly noticed that Arminius and his entire cortege had inexplicably vanished. A short time later, lead elements of the Roman column ran into felled trees and simultaneously came under heavy attack, halting the entire formation (which was several miles long). To make matters worse, a sudden storm deluged them with heavy rain and wind, bogging down chariots and wagons alike. Without warning, the entire column was then assaulted from both sides, along its entire length. Screaming tribesmen, hurling javelins, rushed upon startled and unprepared Romans. The thick forest and mud made it impossible for Roman units to form up into their usual fighting layouts. The whole thing quickly disintegrated into a disorganized brawl, a German specialty!

A remnant of Varus' men valiantly rallied, congregating in a hastily built fort. There, they gallantly held out for two days, but, in the end, were overwhelmed. Varus himself, embarrassed and disgraced, committed suicide, along with most of his staff. Only a handful of mounted cavalrymen escaped on horseback to tell the story. Everyone else, including all dependents and camp followers were annihilated. A few who were taken captive were later used as pitiable human sacrifices to German gods. The death toll was in excess of 25,000!

Back on the Italian peninsula, Augustus was badly shaken by the news, and all of Rome with him! He had been told just weeks earlier that everything was well in hand. He was understandably shocked to discover that he had been lied to by his (now deceased) relative. New legions were dispatched to the area and fighting continued. As the Empire imploded in the later years, Germanica was among the first areas to fall from Roman control, although persistent internal squabbling prevented any effective strategic unity of Germanic tribes against the Romans.

The Teutoburger Wald battlefield laid undisturbed for six years. When Roman units finally found it in 15AD, they discovered only skeletal remains of their comrades. They paused long enough to provide all they could find with a decent burial. Shortly thereafter, the Roman Commander, Germanicus Caesar, nearly suffered the same fate as had Varus! Another ambush, again engineered by the same Arminius, was almost successful. This time, Germanicus was better prepared than had been Varus. He successfully withdrew, salvaging most of his force.

Arminius continued his impassioned (and ultimately doomed) defense of his homeland for another five years, only to be murdered by kinsmen as a result of internal, political intrigue. He never saw his fortieth birthday. His son grew up in Rome, having never known his father. Like Scotland, Germanica was never completely pacified.

Lessons:

Any commander can be forgiven for being defeated. No commander will ever be forgiven for being surprised. Complacency is the deadly enemy of all warriors. The worst attack is always a surprise attack. Be taken in at your peril!

Historians in the next century may say of our time that the greatest mistake made in the wake of WWII was allowing Germany to reunite. Even today, a reunited Germany is surrounded by seven, nervous neighbors! The "old stone gods" may indeed leap to life once more.

Passionate ideologues rarely die of old age!

Glossary

+P – ammunition with more than the normal allowable chamber pressure for a caliber as defined by SAAMI (see below). This is often an attempt to provide more potent ammunition for an older caliber to extend its useful life.

+P+ – ammunition that has even more powder than +P, but still considered safe for strong handguns, and sill regulated by SAMI (see below),

1911 – the designation for the US military standard pistol made by Colt and used until the 1980's. Now made by many manufacturers and still a favorite of shooters.

22 – usually refers to 22 caliber rimfire ammunition and the firearms made to shoot it. Not suitable for self-defense but usually used for practice and informal shooting.

223 Rem – the civilian designation for 5.56x45 mm ammunition.

308 Win – the civilian designation for 7.62x51 mm ammunition.

30 Soviet – a variant designation for 7.62x39 mm ammunition.

45 ACP – 45 caliber Automatic Colt Pistol, a cartridge made popular by the 1911 pistol.

5.56x45mm – the military designation of the 223 Remington caliber ammunition used by the M16, AR-15 and other military rifles.

7.62x51mm – the military designation of the 308 Winchester caliber ammunition used in the M14 and other military rifles.

7.62x39mm – the Soviet military designation of the 30-caliber ammunition usually associated with the AK-47 or SKS rifles.

AD – accidental discharge; also, unintentional discharge (UD) or negligent discharge (ND). Describes any time that a firearm discharges a round of ammunition without intent on the user's part.

Aftermarket – a part or accessory added to a gun that is not originally part of the gun as provided by the manufacturer.

AK-47 – a Soviet-designed automatic rifle used extensively around the world. It is cheap to make and very reliable.

AR-15 – the semi-automatic version of the military M16, used frequently by police and civilians.

AR – assault rifle; originally referred to AR-15's and M16's and their descendants. Now often used generally for all evil-looking black rifles.

Autoloading – a type of firearm that automatically loads a new round of ammunition into the chamber each time it fires.

BATF[E] – Bureau of Alcohol, Tobacco and Firearms; the federal agency tasked with regulating firearms in the US. Explosives were added to the name recently.

CCW – concealed carry weapon, usually used to describe the state permit allowing a person to carry a firearm for self-defense.

Centerfire – a cartridge with a primer in the center of the base instead of around the edge (rimfire). All modern self-defense cartridges are centerfire.

CMP – Civilian Marksmanship Program; a federal program to sell used semi-automatic military firearms to civilians for training purposes.

CPD – Chicago Police Department.

DAO – Double Action Only; an autoloading pistol, or revolver, that can only be fired via the trigger-cocking mode. The hammer cannot be manually cocked, nor in the case of the autoloader, does the hammer remain in the cocked position when the slide reciprocates.

Double-action – an autoloading pistol, or revolver, that can be fired either

Glossary

via the trigger-cocking mode, or the manual-cocking mode.

DOA – dead on arrival; said of a person who dies before arriving at a hospital.

DRT – dead right there; said of a person who dies at the scene.

DT – defensive tactics; refers to unarmed hand-to-hand tactics.

EDP – emotionally disturbed person

ER – emergency room.

FDR – Farnam Defensive Rifle, an M-1 rifle with a forward mounted (scout) scope.

FN – Fabrique National, a firearms manufacturer in Belgium, known for the FAL rifle and the Browning Hi-Power pistol.

Gr – grains; the unit of measurement for powder and bullets. There are 7000 grains to a pound.

Half-cock – an intermediate position of a pistol or revolver hammer where the hammer is neither cocked for firing or fuly forward resting on the firing pin, usually found on older guns as a safety mechanism.

Hollowpoint – a bullet with a hole in the nose to allow for expansion upon impact.

IWB – inside the waistband holster, used for concealment under a jacket.

KB – kaboom, said of pistols that are damaged or destroyed while firing. Typical causes are an obstructed barrel or defective ammunition.

LAPD – Los Angeles Police Department.

LE – law enforcement.

LEO – law enforcement officer.

LPVO – low-power, variable optics

M14 – a full-automatic military rifle firing 7.62 mm ammunition used in the 1950's and early 1960's.

M16 – a full-automatic military rifle firing 5.56 mm ammunition designed in the 1960's that is the predecessor of the current US military rifles.

Muzzle brake – a device added to the muzzle of a rifle to reduce felt recoil.

NATO – North Atlantic Treaty Organization, a military alliance of European and North American countries.

NRA – National Rifle Association.

N – frame - a large revolver frame from Smith & Wesson, commonly used for 44 Magnums. S&W frame sizes from smallest to largest: J, K, L, N, X.

NJ – New Jersey.

NJSP – New Jersey State Police.

NJIT – New Jersey Institute of Technology.

NY – New York.

NYPD – New York Police Department.

NY triggers – a Glock factory replacement trigger mechanism that increases the pistol's trigger pull weight.

OC – oleoresin capsicum, the active ingredient in pepper spray.

PD – police department.

PI – private investigator.

Pow'RBall – a type of defensive ammunition manufactured by Cor-Bon.

RA – Robinson Arms, a firearm manufacturer

RDO – Red Dot Optic, a non-magnifying optical sight for rifles, pistols, and shotguns, featuring a single red dot (or circle, or some combination) as an aiming point.

RO – range officer.

Glossary

RPG – Rocket propelled grenade

S&B – Sellier & Bellot, a foreign ammunition manufacturer.

S&W – Smith & Wesson, a firearm manufacturer.

SA – either Springfield Armory (a firearm manufacturer) or South Africa.

SAAMI – Sporting Arms and Ammunition Manufacturer's Institute, an association of manufacturers that defines voluntary standards for firearm and ammunition safety.

Single-action – a pistol or revolver that does not feature a trigger-cocking mode. The hammer must be first manually cocked, or in the case of the 1911 pistol, the hammer is always cocked as the pistol is carried.

SKS – a predecessor to the AK-47 assault rifle.

SO – sheriff's office.

TiAlN – titanium aluminum nitride – a bronzing coating for steel.

TRB – tap, rack, bang; the sequence of actions suggested on detecting a semiautomatic pistol malfunction. This is still used as an abbreviation, although the "bang" has been replaced by a more cautious verb, such as "resume."

Index

1

1903 Springfield 109
1911 ... 24, 101, 105, 129, 130, 131, 133, 134, 135, 143, 146, 147, 148, 176, 177, 267, 271
1917 American Enfield 109

3

300H&H 142, 191
303Br SMLE Enfield 138
357SIG ... 46, 85, 86, 92, 94, 97, 98, 99, 146, 181

4

40S&W ... 3, 86, 91, 94, 97, 99, 100, 130, 131, 145, 146, 148, 153, 181
44Mg Ruger 138
45 GAP 94
458WM 142
45ACP 24, 71, 86, 93, 99, 129, 131, 146, 147, 148, 169, 212

5

58WinMag 142

9

9mm ... 3, 27, 76, 77, 78, 85, 93, 96, 98, 99, 129, 130, 131, 145, 146, 147, 165, 181, 190

A

Accidental discharges 57
Adrianople, Battle of 256
Adrianople, Turkey 253
Advanced Training Systems 133
Aetius 257
Aguila IQ 85
AK-47 .... 107, 176, 267, 268, 271
Alcatraz Federal Prison 212
AN-94 107
AR-15107, 110, 111, 114, 119, 121, 267, 268
Arizona 6
Arminius of the Cherusci 263
Arrowsmith, Kelee 39
Articles of Confederation 214
Ashley, John 215
ASTA 46, 47, 48
Atlanta, GA 59
Attila the Hun 254, 257
Augustus 263
Automobiles 50, 52

B

Balangiga, Massacre at 229
Baltimore, MD 3, 208
Beamhit 132
Benelli Super-90 138
Beretta 132
Beretta 92 .... 74, 92, 97, 101, 103, 144, 145, 147, 149, 202
Beretta CX4 132
Beretta Tomcat 95
Bill of Rights 216
Boers 233
Boone, Daniel 208
Boston, MA 214
Bowdoin, James 214, 215
Bowling alley 21
Bradley, Omar 226
Bratton, William J. 74, 143, 144, 145, 149
Brooklyn, Battle of 211
Bryco 9mm 77
Buckshot 86, 123, 124, 194
Bulge, Battle of the 226
Bunker Hill 207
Burnside, Ambrose 220, 221
Byzantine 245

C

Capetown, South Africa .. 28, 39, 61, 144, 188, 190, 198

273

Capone, Al 212
CAR-15 113
CBC shotguns 194
CCW 15, 47, 149, 180
CETME 120
Charlemagne (Charles the Great)
  256
Chicago Police Department .. 23, 24, 212, 268
Chicago, IL23, 24, 57, 71, 211, 212, 268
Chudwin, Jeff 119
Churchill, John, Duke of
  Marlborough 241
Churchill, Winston 100
Claremont, France 243
Cold Steel 39, 143
Cold Steel City Stick 25
Cold Steel Culloden 47
Cold Steel Ti-Lite 47
Cold Steel Vaquero Grande ... 47, 49
Colley, George 237
Colt Car-15 122
Comnenus, Alexius 244
Connell, Thomas 229
Constantine the Great 243
Constantinople .. 243, 247, 248, 256
Constitutional Convention 217
Cor-Bon 37, 85, 131, 138, 182, 186, 270
Cor-Bon 115grHP 47
Cor-Bon 125grHP 47
Corner Shot 132
Cornwallis, Charles 211
Crusades 243, 247
CZ 75 .... 76, 101, 186
Czolgosz, Leon 230

D

D Ring 119, 120
da Vinci, Leonardo 141
Dandolo, Enrico 247
DAO 100, 268
DaSilva, Joe 141, 190
Defiant Munitions 37
Denver, CO 67

Detroit, MI 71
DRT 59, 73, 74, 75, 78, 190, 269
Duty Gear 143

E

EDP 76, 269
EO Tech 133

F

Falaise Pocket 226
Federal American Eagle 36
Felter, Brian 132
Flashlight 23, 39, 40, 90, 194
Flechette 86, 163
Fourteenth Amendment 11
Fredericksburg, Battle of 220

G

G32 .... 46, 49
Gage, Thomas 208
Garand 35, 109, 110, 112, 183
Geoffroy de Villehardouin 248
George I 242
George III 209
Germanica 263
Germanicus Caesar 261
Gerry, Elbridge 216
Gettysburg, Battle of 217
Gettysburg, PA 217, 221, 257
Ghent, Treaty of 208
Givens, Tom 32, 50, 198
Glock 34, 57, 68, 69, 70, 85, 89, 91, 92, 93, 94, 97, 98, 99, 100, 103, 104, 105, 130, 131, 132, 133, 143, 144, 145, 146, 149, 182, 194, 270
Glock 17 .... 97, 182
Glock 18 182
Glock 19 27, 75, 92, 93, 97, 98, 182
Glock 20 138
Glock 21 .... 75, 93, 105, 106, 134
Glock 2257, 79, 86, 89, 90, 94, 144, 153
Glock 23 61, 73, 89, 93, 97, 99, 144
Glock 26 .... 103, 104
Glock 32 .... 47, 85, 93, 97, 98

Index

Glock 37 .... 93, 94
Gold Dot 57, 61, 73, 77, 79, 94
Grant, Ulysses S 221
Gratian 254
Great Treck, the 233

H

H&K G3 120
H&K MP5 120
Hamilton, Alexander 211
Hamlin, Perez 215
Harries Technique 39
Harrisburg, PA 46
Heine, Heinrich 261
Hitler, Adolf 225, 227
Hobbs, Leland 226
Hooker, Joe 221
Horns of Hattin, Battle of the 245
Houston, TX 24
Human, Mark 39
Hunting 137
Hussein, Saddam 160, 162

I

IALEFI 104, 130, 151, 154
IMI Galil 152
Innocent III 247
Iraq ... 120, 159, 160, 162, 163, 166, 167, 169, 171, 184, 250
Iron sights 20, 114, 116, 138
Istanbul 243, 256

J

Jackson, Andrew 208
Jackson, Tom 221
Jackson, Tom "Stonewall" 218
Jerusalem 244
Joubert, Peit 237
JSAP 147

K

Kahr 9mm 27
Kahr P9 96
Kapmes 104
Kel-Tec 95
Kel-Tec 380 96

Kel-Tec P32 95
Kevlar 149
Krebs Custom 111
K-trigger 91, 92
Kuwait 159
KYDEX 33, 194

L

Lasermax 133
Le Prestre, Sébastien, Marquis of Vauban 239
Lee, Robert E 221
Lee, Robert E. 217
LEO 3, 29, 34, 63, 65, 66, 73, 74, 77, 85, 89, 93, 94, 98, 103, 154, 190, 269
Less-lethal force 70
Lexington Common 209
Lincoln, Abraham 220, 242
Lincoln, Benjamin 215
Lizard Brain 8
Los Angeles, CA 27, 51, 149
LAPD 73, 143, 144, 145, 269
Louis XIV 239, 242
LPVO 114, 138, 269

M

M1 Garand 109
M16 .... 87, 112, 120, 121, 164, 167, 175, 178, 183, 267, 268, 270
M1A 19, 110, 116, 179
M4 .... 107, 110, 119, 120, 164, 169, 171, 175
M70 19
M855 164
M9 .... 147, 148, 172
Maccabeus, Judas 262
Machiavelli 22
Madison, James 208, 216
Majuba Hill 238
Manassas (Bull Run), Battle of .. 220
Manzikert, Battle of 244
Mao Zhe-dung 259
Massachusetts 214
McDowell, Irwin 220
McGurn, "Machinegun" Jack 212

275

McKinley, William 230
Meade, George 219, 221, 257
Memphis, TN 32
Molotov cocktail 50
Monongahela, Battle of 208
Montgomery, Bernard 225
Moran, George "Bugs" 212
Mormon Church 19
Mossberg 590 .... 127, 178
Mossberg 590A1 .... 125, 126
Mossberg M590A1 125
Murder-suicide 65
Mussolini, Benito 227

N

NAA 95
Nakma Chu, India 258
Nedao, Battle of 257
Nehru, Jawaharlal 258
New Jersey 4, 64, 208
New Jersey Institute of Technology 4
NJSP 64
Normandy Invasion 225
North Dakota 41
NRA 223
NTI 46, 50
NY Trigger 103, 104, 182

O

OC spray 39, 40, 41, 51, 52, 53, 66, 68, 270
Old Testament 6, 17
Operation Luttich 225
Orange Free State 236
Osman, the first sultan 244

P

P99 131
Parker, John 209
Patton, George S. 225
Philippines 175, 229
Pickett's Charge 217
Pitcarin, John 209
Pleasants, Henry 221
Poker 5, 239
Pope, John 221

Prescott, William 210
Pretorius, Andrius 234
Prince Eugene of Savoy 241
Putnam, Israel 210

R

Rabbit Chasing 5
Raven 25 71
Raven 25ACP 60
Raven Arms 60
Rawlinson's Charge 217
RDO 125, 270
Red dots 114, 175
Remington 870 .... 59, 124, 126, 194
Remmington M7600 132
Richard the Lionhearted 247
ROBAR 89
Robinson Arms .. 110, 111, 179, 270
Roman Catholic Church 249
Romulus Augustulus 255
Roosevelt, Theodore 230
Rottweil semiauto shotgun 194
RPG 50, 271
Ruark, Bob 140
Ruger 4, 74, 99, 121, 132, 138, 139
Ruger Mini 30 121
Ruger Mini-14 65

S

S&W 4516 100
S&W M10 10
S&W M340PD 46
S-4 33
SAAMI 271
Sabre 40
Saladin 245, 247
Saratoga, Battle of 207
Saudi Arabia 10, 203
Schwimmer, Reinhardt H 211
Scope . 19, 113, 114, 116, 138, 142, 191, 192
Seat belt 50
Second Amendment 12, 16, 213, 217
Second Chance 149
Shays, Daniel 214

Shays's Rebellion 213
Shepard, William 215
SIG Sauer 146
SIG Sauer 320 92
SIG Sauer P220 77
SIG Sauer P229 .... 91, 100
SIG Sauer P239 27
SIGMA 130
SigPro 90
Simmunition drill 46, 66
SKS 110, 267, 271
Smith & Wesson
    1911 104
Smith & Wesson 1911 90
Smith & Wesson 340PD 98
Smith, Jacob "Howling Jake" 231
SMLE Enfield 48
snubby 37, 51
Snubby 47
Somme River, Battle of the 217
South Africa 28, 29, 52, 61, 63, 185,
    198, 233
Spanish-American War 229
Speer 94
Springfield Armory 110, 179, 271
Stealth Existence 14, 15, 44
    stealth failure 15
Streamlight 40
Surefire 194

T

Tac-Con 50
Taurus 95
Taurus 38Spl 98
Taurus PT92 .... 101, 103
Tecumseh 207
Terrorism 201
Teutoburger Wald 261
TiAIN 133, 134, 271
Townshend, Sir Charles 250
TRB 105, 271

TSA 24
Tueller Drill 154

U

United Kingdom 64
University of Michigan Law School
    11
Urban II 243, 244
USCG 112
USMC 147, 162

V

Van der Goltz, Colmar 251
Varus, Publius Quintilius 263
Vicksburg, MS 221
Victory disease 233, 239
Vietnam .... iv, 14, 31, 120, 122, 147,
    164, 165, 168, 169, 170
Voortrekkers 234

W

Walther P99 101
Walther PPK 194
Walther PPK/S 97
War of Spanish Succession 239
Washington, DC 34, 208
Washington, George . 207, 208, 211,
    214, 216, 217
Wayne, Anthony 207
West Palm Beach, FL 60
Westley-Richards breechloader .. 236
Whisky Rebellion 217
Winchester 95, 145, 191, 267
Wisconsin 15, 63, 85
Wyoming 67

Z

Zara, Hungary 248
Zulu 234
Zylon 149

Books Available from Defense Training International

The Farnam Method of Defensive Shotgun and Rifle Shooting
by John S Farnam, 1998

The Farnam Method of Defensive Handgunning (Second Edition)
by John S Farnam, 1998

Teaching Women to Shoot: A Law Enforcement Instructor's Guide
by Vicki Farnam and Diane Nicholl, 2002

Women Learning to Shoot: A Guide for Law Enforcement Officers
by Diane Nicholl and Vicki Farnam, 2006

FlexCCarry™ Solutions: A Positive Guide to Off-Body Carry
by Vicki Farnam, 2024

Guns & Warriors: DTI Quips, Volume 1
by John S Farnam, 2006

Guns & Warriors: DTI Quips, Volume 2
by John S Farnam, 2025

For information on training by John and Vicki Farnam,
see their web page, **defense-training.com**

Defense Training International, LLC
John S Farnam
1281 E Magnolia St, D339
Ft Collins, CO 80523

www.ingramcontent.com/pod-product-compliance
Lightning Source LLC
Chambersburg PA
CBHW060453030426
42337CB00015B/1577